NATURE OF INFORMATION

This book redefines our understanding of informati data or communication—it presents information fabric of reality itself. Drawing from disciplines a physics, evolutionary biology, cognitive science, arti metaphysics, the book explores how information o organising force that sculpts the cosmos, consciousn delves into how biological evolution is driven by inforn neural complexity gives rise to consciousness, and how mirror the architecture of thought itself. The narra theories—like entropy and natural selection—to mod integrated information theory, quantum computation, intelligence. By weaving scientific analysis with ph the book argues that information is not a passive n force, guiding evolution, shaping reality, and potent fifth fundamental force of nature. This book will be i researchers, and general readers across disciplines such physics, evolutionary biology, computer science, neuro mind, and consciousness studies. It will also serve as a for thinkers, technologists, and educators seeking informational scaffolding behind nature, mind, and ma

Joram Beda (Dr.) is an Indian Administrative Service graduate from JIPMER, currently with the Governme interests include climate resilience, renewable energy His recent publication is "People's Perceptions, Practi Knowledge on Climate Change in Meghalaya" (2025).

NATURE OF INFORMATION

Shaping Reality, Life, and Consciousness

Joram Beda

Routledge
Taylor & Francis Group
LONDON AND NEW YORK

Designed cover image: Getty

First published 2026
by Routledge
4 Park Square, Milton Park, Abingdon, Oxon OX14 4RN

and by Routledge
605 Third Avenue, New York, NY 10158

Routledge is an imprint of the Taylor & Francis Group, an informa business

British Library Cataloguing-in-Publication Data
A catalogue record for this book is available from the British Library

ISBN: 978-1-041-24224-6 (hbk)
ISBN: 978-1-041-24225-3 (pbk)
ISBN: 978-1-003-74076-6 (ebk)

DOI: 10.4324/9781003740766

Typeset in Sabon
by Apex CoVantage, LLC

To my lovely children, Mamko and Aimee—my brightest lights—and to Yapu, the heart that holds us together.

CONTENTS

ILLUSTRATIONS

PREFACE

Why is it that so much of what we imagine, wish for, or sketch in the mind becomes real once we set ourselves to it? The house first drawn on paper, the violin carved from rough wood, the microchip refined from sand, the spacecraft designed from equations—each began as a spark in someone's brain. It starts as a flicker, becomes a picture, takes shape as a plan, and is finally crafted for use.

From time immemorial, philosophy and science have asked: What is a thought made of? At the cellular scale, thought is electrochemical signalling in neural circuits: ions pass through voltage- and ligand-gated channels, generating all-or-none action potentials that travel along axons; at synapses, neurotransmitters diffuse across the cleft and bind receptors, producing postsynaptic currents; with experience, synapses and dendrites remodel—their efficacies strengthened or weakened through plasticity. But how do such microscopic events compose a painting that awes the whole world? How do they yield equations that predict the behaviour of a distant black hole or discover the curve of an airfoil that lifts a jumbo jet into the sky? The brain does this by recruiting vast, distributed networks whose activity is sculpted by inhibition, neuromodulators, and dendritic computation, supported by glia, and trained by error signals. These networks learn internal models, plan, and translate intention into motor programmes; those programmes drive muscles and tools, turning information patterns in the brain into organised work in the world.

Perhaps the answer begins with a simple observation: life runs on stored patterns. DNA—life's script—encodes four nucleotides (A, T, G, C). Arranged into sequences, varied by mutation, and sifted by natural selection, they store aeons of information about survivability. Through the central dogma

(DNA → RNA → protein), ribosomes spin amino acid chains that fold into proteins—ion channels and pumps, enzymes, motors, photoreceptors, collagen, claws and wings—sophisticated macromolecular machines that give matter form and function.

As organisms grew more complex, they evolved a second archive: nervous systems. If the genome is a long memory of what worked across deep time (phylogenetic information), neural tissue is a fast, flexible memory of what works now (ontogenetic and cultural information). Brains record regularities in the environment—what to eat and avoid, how to farm, how to evade predators, how to cooperate. Culture extends this memory across individuals and generations—stories, techniques, laws, and tools as externalised synapses binding communities together. Together, the genomic and the neural systems multiply an organism's chances of survival.

Yet their characters differ. Genomic rollout is directional—a disciplined flow from nucleic acid to protein (with known exceptions like reverse transcription), constrained by biochemistry. Neural computation is exploratory—predictive, plastic, generative. And the human brain—an extravagant flowering of cortex and connectivity—not only adapts to the world but imagines alternatives and makes them real. It recombines models, simulates futures, and acts through hands and institutions to test those futures against reality.

Is information, then, merely a description we write after the fact—or is it something more basic? Working postulate: information is not just bookkeeping about matter and energy but a causal, organising principle that constrains and enables them. The so-called universal constants can be read as the world's boundary conditions—numerical rails along which matter must run, and models must fit. This does not prove that reality is information at root, but it suggests that persistent patterns and lawful limits are more than labels; they are scaffolds on which both matter and meaning reliably hang.

Consider convergent evolution. Under similar constraints, life repeatedly discovers similar solutions: camera eyes in vertebrates and cephalopods; wings in insects, birds, and bats; torpedo shapes in ichthyosaurs, tuna, and dolphins. Different lineages, shared informational landscapes—fluid dynamics, optics, load-bearing geometry—pull designs into the same basins of attraction.

Or consider fractal branching, where flow meets limit and purpose. River deltas, bronchial trees, and dendritic arbours ramify with self-similar geometry; roots beneath soil and roads above it solve the same transport problem with related architectures. In nervous systems, it is not that sodium ions somehow "travel" from cortex to the limb as little couriers of will; rather, patterns of activity propagate through circuits to motor neurons, which release neurotransmitters at the neuromuscular junction to contract muscle. In governance, too, nothing material need relocate for change to occur; what moves are messages—structures of information—that reshape behaviour: different substances, similar constraints, the same families of shape. It is as if matter and meaning prefer the same channels when guided by flow, limit, and goal.

Across evolution and behaviour, information behaves like a real cause: it selects, shapes, constrains, and enables. We have always offloaded our burdens into tools. Stones broke nuts; fire softened proteins; writing moved memory beyond the hippocampus into clay, paper, discs, and clouds; engines put muscle outside the bone. And now, with artificial intelligence, we begin to augment not only memory and muscle but modelling—pattern-finding and decision-making themselves. This is not mere convenience. It answers an older impulse: the information architecture within us seeking expression beyond us—an extracorporeal life, a second nervous system braided into ours.

Put simply, imagination becomes real because brains are engines that convert information into work under constraint. DNA and neurons are its great looms; culture and computation are its newer mills. The clearer we see the constraints, flows, and purposes that govern this thread, the better we can understand why thoughts so often become things—and how to guide that transformation toward lives and worlds worth making.

ACKNOWLEDGEMENTS

To **Lord Shiva** for grace, and to **Yapu**, **Mamko**, and **Aimee** for love and patience—my deepest thanks.

GLOSSARY AND ABBREVIATIONS

Glossary

Amazon Web Services (AWS): A cloud computing platform providing on-demand compute, storage, and other services.

Arithmetic–logic unit (ALU): The part of a CPU that performs arithmetic and logical operations. Used metaphorically in the book to compare biochemical decision steps (e.g., PTC).

ARNTL/BMAL1: A core circadian transcription factor. Heterodimerizes with CLOCK or NPAS2 to drive rhythmic gene expression via E-box elements.

CLOCK: A circadian transcription factor (name from "Circadian Locomotor Output Cycles Kaput"). Forms a heterodimer with BMAL1/ARNTL.

Cryptochrome (CRY1/CRY2): Flavin-containing proteins that constitute the negative-feedback arm of the circadian clock, inhibiting CLOCK–BMAL1 activity.

Extracorporeal membrane oxygenation (ECMO): A life-support technique in which blood is oxygenated outside the body and returned to the circulation.

Fenna–Matthews–Olson (FMO) complex: A bacterial photosynthetic pigment–protein complex often discussed in quantum biology for exciton transport.

General Data Protection Regulation (GDPR): The European Union regulation governing data protection and privacy and the transfer of personal data.

Global workspace theory (GWT): A cognitive theory positing that information becomes conscious when broadcast to a global workspace accessible by many processes.

Integrated information theory (IIT): A theory proposing that consciousness corresponds to the amount and structure of integrated information (Φ) generated by a system.

NASA/NACA: NASA is the U.S. civil space agency; NACA was its aeronautical research predecessor.

Natural language processing (NLP): A field of AI focused on the computational analysis and generation of human language.

Period proteins (PER1/2/3): Clock proteins that accumulate and, with CRYs, inhibit their own transcription by repressing CLOCK–BMAL1.

Perturbational complexity index (PCI): An empirical index of consciousness based on the complexity of EEG responses after perturbation (typically TMS).

Temperature-Dependent Sex Determination (TSD): A mechanism where incubation temperature determines the sex of the developing embryo (common in many reptiles).

University College London (UCL): A public research university in London, referenced for research discussed in the book.

Abbreviations

ALU	arithmetic–logic unit
ARNTL	Aryl hydrocarbon receptor nuclear translocator-like 1 (BMAL1)
AWS	Amazon Web Services
BMAL1	Brain and muscle ARNT-like 1 (gene symbol ARNTL)
CLOCK	Circadian Locomotor Output Cycles Kaput
CRY	Cryptochrome proteins (CRY1/CRY2)
DNA	Deoxyribonucleic acid
ECMO	Extracorporeal membrane oxygenation
FMO	Fenna–Matthews–Olson complex
GDPR	General Data Protection Regulation
GWT	Global workspace theory
IIT	Integrated information theory
NACA	National Advisory Committee for Aeronautics
NASA	National Aeronautics and Space Administration
NLP	Natural language processing
PCI	Perturbational complexity index
PER	Period proteins (PER1/PER2/PER3)
RNA	Ribonucleic acid
SSC	Spitzer Science Centre
TSD	Temperature-dependent sex determination
UCL	University College London

1

FROM EQUATION TO ARSENAL

How $E = mc^2$ Recast World Power

1.1 The Equation That Changed the World

By early 1945, after six years of relentless exhaustion that reached from quiet village squares to the shadowy war rooms of Europe's leaders, the eyes of the world turned towards the Pacific. There, the final and most ferocious phase of the conflict raged on. Hitler was dead, Mussolini hanged, yet Japan—its navy in ruins, its cities aflame—refused to surrender. For four straight years, the Allied Forces enforced a naval blockade and starved the Home Islands of oil and other raw materials; Tokyo's night sky glowed red as American bombardment turned elegantly built wooden neighbourhoods to smouldering ruin. Yet every new Allied landing revealed a determined nation still capable of astonishing resistance against all odds.

At Iwo Jima in February, 6,800 U.S. Marines were killed seizing an airstrip just eight square miles wide; nearly every one of the island's 18,000 Japanese defenders fought to the last cartridge, putting up a befitting resistance. In sharp contrast to many battles in other war theatres, the Japanese defence strategy placed an emphasis on unrelenting resistance, displaying a will to fight to the death rather than surrender. This strong will led to an incredibly high death rate, with almost all of the Japanese troops on the island dying. In April, the invasion of Okinawa began with the loud firing of 1,900 naval guns. During the gruelling 11-week war that followed, 12,000 American soldiers, more than 100,000 Japanese troops, and maybe as many as 150,000 civilians would die, leaving the island scarred and ruined. Okinawan civilians, driven by fear and desperation, chose death over captivity—entire families tragically leaping from the island's rugged limestone cliffs rather than falling into enemy hands. Admiral Raymond Spruance grimly described the

DOI: 10.4324/9781003740766-1

ferocious battle for Okinawa as "a preview of what the invasion of Japan's Home Islands would entail."

Allied planners were already carefully planning Operation Downfall, an ambitious, two-part assault that would end World War II for good by taking over the Japanese mainland, even while the battle was still going on. Military planners thought that an attack on Japan itself could inflict up to one million Allied casualties. Meanwhile, projected Japanese losses climbed to unimaginable numbers, underscoring the horrifying scale of devastation such an invasion would bring. In Japan, preparations grew increasingly desperate: waves of kamikaze aircraft multiplied, their pilots committed to suicide missions aimed at crippling Allied ships; schoolchildren, some as young as 14, practised defence drills wielding bamboo spears, readying themselves for a final, hopeless stand; and remnants of the Imperial Navy dispatched their last operational warships on one-way voyages, their fuel supplies insufficient for a return journey. As late as the first week of August 1945, the trajectory of the war still pointed inexorably towards a bloodbath on the beaches of Kyūshū.

Just after eight in the morning of 6 August, a young office worker in Hiroshima gestured for his daughter to pose beside the family's new bicycle on a quiet side street off Nakajima-dōri. Kneeling down, he adjusted the focus on his boxy Vest Pocket Kodak and pressed the shutter. Mid-click the eastern sky detonated. A magnesium-white flash over the city's T-shaped Aioi Bridge vaporised the camera's film emulsion; seconds later a blast front—hotter than a steel furnace roared across the streets of Hiroshima, reducing everything in its path to dust and devastation. The officer, daughter, the bicycle, and the asphalt beneath them were gone—reduced to a negative shadow etched into stone. Seconds earlier, at 08:15 AM, the Boeing B-29 *Enola Gay* had released a four-ton uranium bomb nicknamed "Little Boy" from 31,000 feet. After 43 seconds of free fall, the bomb exploded 580 metres above the Shima Surgical Clinic, where a fission chain reaction converted seven-tenths of a gram of matter into 15,000 tonnes of TNT equivalent explosion. Tens of thousands died almost instantly, before they could even hear the sound of the explosion, and five square miles of bustling city vanished in an ascending column of flame.

Three days later, at 11:02 AM on 9 August, a plutonium device "Fat Man" exploded above Nagasaki's Urakami valley. The surrounding hills focused the blast like a trumpet bell; factories that had built torpedoes for Pearl Harbor collapsed in a single groan of twisting steel. Thirty-nine thousand people died instantly; radiation sickness and burns would double that figure by winter. In both cities survivors reported a silent white light "hotter than the sun" followed by a darkness so complete it seemed the world had come to an end.

Within hours, Tokyo decoded intercepted American broadcasts confirming that only a single aircraft had been needed for each annihilation. The

FIGURE 1.1 View of a Hiroshima bridge 4,400 feet east of ground zero; aftermath of the 6 August 6 1945 atomic bomb blast in Hiroshima; Official U.S. Army photograph. Library of Congress, *New York World-Telegram* and *Sun* Collection. Public domain.

same night, news arrived that Soviet armies had poured into Manchuria, ripping away the last hope of a mediated surrender. The Japanese Supreme Council convened under palace eaves while incendiaries still smouldered in half the nation's prefectures. For the first time in history, the Emperor spoke across the clashing opinions of his ministers: "We must endure the unendurable." Just past midnight on 15 August, Hirohito's recorded voice reached a stunned populace, announcing the cessation of hostilities and citing "a new and most cruel bomb." Japan surrendered. Two years later Japan's new constitution would renounce "war as a sovereign right" and prohibit "the threat or use of force forever."

1.2 From Ruins to Rivalry: The Postwar World Reconfigured

The end of World War II in 1945 ushered in a dramatic reconfiguration of global powers. The old imperial powers—Britain, France, Germany, and Japan—emerged either devastated or diminished, their colonies moving towards independence and their economies in ruins. In their place, two states emerged as dominant powers of a new world order: the United States and the Soviet Union. These two nations, ideologically opposed and armed with unprecedented military and economic might, began to shape what would become a bipolar international order.

The United States emerged from the war with its mainland largely intact, its economy surging, and its military apparatus global in reach. By 1946, it accounted for approximately 60% of the world's industrial output, possessed two-thirds of its gold reserves, and enjoyed a monopoly on nuclear weapons. In contrast, the Soviet Union had suffered colossal losses in both men and material—over 20 million dead and widespread destruction—but its Red Army occupied Eastern Europe, and its wartime mobilisation had turned it into a formidable industrial and military power. The Red Army occupied Eastern Europe, transforming the region into a buffer zone of satellite states that not only extended Soviet influence on the heart of the European continent but also provided a critical workforce, industrial capacity, and raw materials to fuel its postwar recovery and military expansion.

As tensions between the former Allies intensified, most nations found themselves pulled into one of two emerging geopolitical camps. In 1949, the United States formalised its military and political leadership over Western Europe through the creation of North Atlantic Treaty Organization (NATO)—a collective security pact premised on mutual defence. Twelve founding members, including the United States, the United Kingdom, France, and Italy, pledged that an attack on any one of the member nations would be considered as an attack on all. In response, the Soviet Union established its own security alliance in 1955: the Warsaw Pact, comprising the USSR and seven Eastern Bloc nations, including Poland, East Germany, Hungary, and Czechoslovakia. These two alliances crystallised the global divide between capitalist democracies and communist regimes.

Alongside these military alignments, both superpowers undertook massive programmes to develop and stockpile nuclear weapons. The United States had demonstrated the terrifying potential of atomic power in Hiroshima and Nagasaki and initially held a commanding lead. However, in 1949, the Soviet Union detonated its first atomic bomb, ending the U.S. monopoly. The arms race escalated rapidly. In 1952, the United States tested the first hydrogen bomb, a thermonuclear device far more powerful than earlier atomic bombs. The Soviet Union followed suit in 1953.

By the early 1960s, both nations had deployed intercontinental ballistic missiles, submarine-launched ballistic missiles, and strategic bombers—creating what came to be known as the nuclear triad. Each side amassed thousands of warheads, with the United States reaching over 30,000 nuclear weapons by the mid-1960s, and the Soviet Union eventually surpassing that total in the 1970s. This staggering nuclear weapons buildup was governed by the doctrine of mutually assured destruction:

the understanding that any full-scale nuclear exchange would annihilate both the attacker and the defender.

The alignment of global military blocs and the relentless nuclear arms race defined the structure of the Cold War, freezing much of the world into two antagonistic spheres. While open warfare between the superpowers was avoided, regional conflicts, proxy wars, and constant brinkmanship played out beneath the shadow of nuclear deterrence—a dynamic that would shape global geopolitics for decades to come.

The United States and the USSR armed at a pace history had never witnessed: by 1965 they possessed, between them, more than 37,000 strategic warheads, enough to incinerate human civilisation many times over. Yet the very scale of that arsenal imposed a hard ceiling on the pursuit of war. Strategy no longer revolved around conquest or even attrition—it revolved around survival. In that fragile balance, the world found itself teetering, again and again, at the edge of another world war and irreversible catastrophe.

1.3 Deterrence and the Dance of Annihilation: Cold War Logic Unfolds

Few moments embodied this razor-thin line between peace and war more vividly than the showdown at Checkpoint Charlie in October 1961. As the USSR-backed East German authorities began sealing off West Berlin with concrete and barbed wire—denying Western diplomats access to East Berlin—American tanks rolled into the city's centre. They were soon met by Soviet tanks, stopping just metres apart at the Allied checkpoint dividing the sectors. For 16 hours, they faced each other barrel-to-barrel, engines idling, weapons loaded—each side waiting for the other to blink. Berlin, long a Cold War pressure point, became a live fuse. A single shot, even a mechanical misfire, could have spiralled into a Europe-wide conflagration and possibly started the next world war. As the world watched in horror, holding its breath, Soviet and American tanks faced off, motionless and loaded for war. But behind the scenes, discreet diplomatic channels opened. Through backchannel talks between Robert F. Kennedy and Georgi Bolshakov the standoff was quietly defused, and the tanks began to withdraw. Restraint, not rhetoric, carried the day—and the crisis dissolved into a tense, fragile calm.

Yet the confrontation at Checkpoint Charlie, perilous as it was, would soon be overshadowed. Just a year later, in October 1962, the world truly stared into the nuclear abyss. When American reconnaissance aircraft photographed Soviet nuclear missiles under construction in Cuba—just 90 miles from Florida—the United States imposed a naval "quarantine," risking direct military confrontation with Soviet ships en route to the island. For 13 days, the world held its breath as backchannel negotiations flared and faltered, Strategic Air Command stood at DEFCON 2, and nuclear-armed bombers

FIGURE 1.2 The tank confrontation at Checkpoint Charlie, Berlin (October 1961). For 16 hours, U.S. and Soviet armour stood barrel-to-barrel at the sector boundary before both sides withdrew through backchannel diplomacy.

circled the skies. Unknown to most at the time, Soviet submarines near the blockade perimeter carried nuclear torpedoes and had standing orders to use them if attacked. In the end, diplomacy prevailed: Khrushchev agreed to withdraw the missiles in exchange for a U.S. pledge not to invade Cuba and the quiet removal of American Jupiter missiles from Turkey. But the lesson endured: miscalculation—just as much as malice—could end the world.

Paradoxically, the fear of absolute destruction spurred a new era of crisis management and arms control. From that moment onward, the logic of deterrence demanded not only military strength but communication and predictability. Washington and Moscow installed a direct hotline to enable immediate contact between leaders. They began sharing limited data on atmospheric nuclear tests to reduce suspicion and false alarms. Arms control talks—once unthinkable—took root: the Partial Test Ban Treaty of 1963 outlawed above-ground nuclear explosions, while the Nuclear Non-Proliferation Treaty of 1968 sought to halt the spread of nuclear weapons beyond the original five atomic states. The world had entered an era where physics, diplomacy, and existential anxiety were now permanently intertwined.

All of it—the Japanese capitulation, the abrupt end of World War II, realignment of geopolitics post-World War II, emergence of new global powers, the uneasy paradoxical peace between the two superpowers, the global chessboard of proxy wars and political realignments, the relentless ever-present shadow of nuclear catastrophe—can be traced to a deceptively simple equation: $\boldsymbol{E} = mc^2$.

1.4 The Equation That Shaped Empires: $E = mc^2$ as Causal Engine

In his 1905 *Annalen der Physik* research note, "Ist die Trägheit eines Körpers von seinem Energieinhalt abhängig?" Albert Einstein showed that when a body gives off energy as radiation, its mass decreases in direct proportion to the energy released—leading to the idea of a body's rest energy:

$$E_0 = mc^2,$$

linking a body's rest energy E_0 to its invariant mass m through the square of the vacuum speed of light c. Einstein's derivation did not arise from speculation about armaments: it was an austere exercise in ensuring that energy–momentum conservation holds in every inertial frame. He considered a thought experiment in which a body emits radiant energy L symmetrically in opposite directions; observers in relative motion must then measure the body's mass as having decreased by $\frac{L}{c^2}$. The conclusion is inescapable: mass is condensed energy.

Because $c^2 \approx 8.99 \times 10^{16}$ J kg^{-1}, even commonplace objects harbour prodigious energy reservoirs. One gram of matter embodies roughly 9×10^{13} joules—about 90 terajoules—enough to sustain the electrical demand of a city like London for several days or, if released in microseconds, to match the explosive yield of the 1945 fission devices.

For nearly three decades the mass–energy equivalence remained primarily in the domain of astrophysics. Arthur Eddington applied it to stellar lifetime calculations, and by 1938 Hans Bethe was using it to quantify the energy generation in the proton–proton chain that powers main-sequence stars. The trajectory of nuclear physics changed decisively in late December 1938, when Otto Hahn and Fritz Strassmann, working in the radiochemistry laboratories of the Kaiser-Wilhelm-Institute in Berlin-Dahlem, observed an unexpected anomaly in their neutron activation experiments. Astonishingly, uranium samples they had bombarded with slow neutrons produced barium—an element with barely half uranium's mass number and incompatible with the prevailing models of radioactive transmutation. Unsure how to interpret the data, Hahn dispatched a detailed account to his long-time collaborator Lise Meitner, who had recently escaped Nazi Germany and was living in exile in Kungälv, Sweden.

Meitner met her physicist-nephew Otto Frisch on 29 December that year; the two spent the next day reconstructing Hahn's measurements on a pad of graph paper beside a snowy forest path. Treating the nucleus as a charged liquid drop, they calculated the energy released if a uranium nucleus cleaved into two lighter fragments. The mass deficit corresponded to roughly 200 MeV per event—precisely what Einstein's $E_0 = mc^2$ predicted for such a split. The implication was clear: the uranium nucleus had not shed an alpha particle or undergone a minor rearrangement; it had divided into two substantial parts, each recoiling with enormous kinetic energy.

In early January 1939, Frisch returned to Copenhagen, repeated the experiment with a modest ionisation chamber, and observed the intense, quantifiable release of energy predicted by Einstein's equation. Frisch published his findings in a short letter to *Nature*, where he introduced the term "nuclear fission," borrowing a word from cell biology to describe the physical division of the atom. Within weeks, laboratory confirmation arrived from Paris, Rome, and New York, and the mass–energy relation migrated from theoretical elegance to empirical certainty. For the first time, scientists had direct experimental proof that a minute quantity of rest mass could be converted into prodigious amounts of energy—an insight that would reshape both physics and geopolitics in the decade to follow.

1.5 From Thought to Thermonuclear: The Manhattan Project and Its Aftermath

That insight—proving that rest mass could be directly transformed into usable energy—set in motion a chain of events that would rapidly carry Einstein's equation from the realm of theoretical physics into the domain of global strategy. By 1939, it was becoming increasingly clear to a small group of physicists that nuclear fission was not merely a scientific breakthrough but a potentially transformative military technology. Among them was Leo Szilard, a Hungarian immigrant who, having fled fascism in Europe, grasped both the scientific and political consequences of the discovery.

Fearing that Nazi Germany might develop an atomic bomb first, Szilard—together with fellow physicists Eugene Wigner and Edward Teller—sought to alert the U.S. government. Recognising that their warnings might not be taken seriously coming from little-known scientists, they turned to Albert Einstein, whose name carried the necessary weight. In August 1939, Szilard drafted a letter for Einstein's signature to President Franklin D. Roosevelt, warning that uranium could be used to build an extremely powerful new type of bomb and that German scientists were actively working on such research. The letter was delivered via economist Alexander Sachs and eventually reached Roosevelt, who responded cautiously but with enough seriousness to initiate further investigation.

By October 1939, Roosevelt had authorised the creation of the Advisory Committee on Uranium, marking the U.S. government's first formal step into nuclear weapons research. What began as a modest programme of theoretical studies and small-scale laboratory work expanded dramatically after the United States entered World War II. In 1942, under the leadership of physicist J. Robert Oppenheimer and Army engineer General Leslie Groves, the project was consolidated under the Manhattan Engineer District—more commonly known as the Manhattan Project. It soon grew into one of the largest scientific undertakings in history, employing more than 130,000 people across top-secret facilities in Los Alamos (New Mexico), Oak Ridge (Tennessee), and Hanford (Washington).

The first milestone in converting theory into reality came on 2 December 1942, when Enrico Fermi and his team at the University of Chicago successfully initiated the world's first controlled, self-sustaining nuclear chain reaction in Chicago Pile-1—a crude but functional reactor assembled beneath the stands of an abandoned football stadium. Three years later, on 16 July 1945, the first atomic bomb was tested in the New Mexico desert. The Trinity test unleashed a 19-kiloton explosion—equivalent to 19,000 tonnes of TNT—and etched an incandescent confirmation of $E_0 = mc^2$ into the fabric of history. Less than a month later, the bombs dropped on Hiroshima (6 August) and Nagasaki (9 August) demonstrated the full military potential of the technology, leading to Japan's surrender and the end of World War II. In just seven years—from the discovery of nuclear fission—an unassuming mathematical equation became the operative principle behind the most consequential weapons ever devised.

What began as a solitary spark of insight in one mind—through the intertwined efforts of scientists, engineers, politicians, and statesmen—recast the architecture of global power.

Much of it traces back—not to a tectonic upheaval, a cosmic collision, or an extinction-level flood—but to a pattern of thought, born in a single consciousness and rendered into symbols. A chain of reasoning. A sequence of concepts. An architecture of information.

1.6 Ideas, Language, and Human Agency: The Triad of Transformation

How is it possible that an idea—a fleeting surge of electrochemical activity in the human brain—could redirect the course of civilisations? That something intangible, without weight or force in the physical world, might ripple outward, reshaping science, technology, diplomacy, and war?

This is the defining enigma of our age: that information—immaterial yet exquisitely structured—exerts genuine causal power over the material world. Not by pushing or pulling like a stone dropped into water, but by organising how we perceive, order, and ultimately manipulate the physical substrate beneath us. The atomic bomb was more than uranium and intricate engineering—it was the tangible product of an informational revelation, propelled by sheer will and coordinated human effort.

Information does not act alone. They are not autonomous agents floating in the void. Like seeds, they require soil, water, and tending. Like energy, information needs a medium through which to propagate. In the domain of human society, that medium begins as a small ignition in the mind: an idea. Without that human hinge, "information" remains description—never quite becoming a plan, a policy, a structure, or a weapon. The passage from private insight to public reality happens largely through language. Yet language is not the same thing as information. The three—ideas, language, and information—move together, but they do different work.

An idea is a structured burst of cognition, a flash of insight or recognition within a conscious mind. It begins as a subjective experience, shaped by attention, memory, and intuition. Ideas are the most volatile—they flicker and vanish unless stabilised by expression. To endure, they must exit the mind and enter a shared domain. This is the function of language: not merely a conveyor of sounds but a vehicle of information. Language reduces the ineffable into the transmittable, the abstract into the shareable. It carries thought across space and time— pressed into clay, written in ink, stored in bits, rendered in pixels, codified in law, and circulated through shared narratives."

But information is deeper still. It is the structure beneath both idea and language, the underlying distinction that allows something to be known, perceived, or communicated. Where language is the form and idea the intention, information is the pattern that holds them together. In Claude Shannon's sense, it is the resolution of uncertainty. In a metaphysical sense, it is what distinguishes one possible world from another—what makes a configuration real, a system meaningful, a sentence intelligible.

And none of these—idea, language, or information—function without interpretation. Human agency is not optional; it is elemental. A message without a receiver is not information. A code without a reader is inert.

So, to speak of information as causal is not to summon some ghostly essence animating the world. It is to recognise that form, once perceived, acquires force—that a pattern, once interpreted, becomes a potential waiting to act. The power of information is not intrinsic to its symbols but arises in the triangular relationship between pattern, interpreter, and action. An equation does not explode. A theory does not govern. A code does not execute itself. It is the mind that reads it, the hand that enacts it, the machine that runs it, the world that responds—this is how information migrates from symbol to system, from abstract description to embodied consequence.

If a single equation could rewrite geopolitics, if a thought scribbled in solitude can leave a crater in a city, then information must be understood not as a passive record of reality, but as an active force—a force that shapes, commands, predicts, and transforms. This is the essence of information: it is not merely what we know. It is the condition of knowing itself—an organising principle that increasingly reconfigures the fabric of what is real.

It was not uranium alone that exploded in 1945 but the informational capacity of the human mind—to encode, transmit, and act upon symbolic knowledge. It was organised knowledge: mathematics translated into engineering, engineering translated into an operational system, and that system translated into physical force.

But what is information, precisely? At first glance, it seems harmless: a set of symbols carefully inscribed on a page—letters, numbers, operators forming a terse formula. Yet when Einstein wrote those symbols, he did more than describe the world; he unveiled a structural pattern woven into the very fabric of the universe.

Information is no inert thing. It wields causality.

Unlike mass or energy, it cannot be weighed or measured by conventional instruments. But its effects? They are etched deeply into history.

It resides in codes, in creeds, and in constitutions. Like a virus, it requires a host—our minds—and culture becomes its conduit. Once accepted, it reshapes perception, and by extension, reality itself.

Ideas spread not just because they are true but because they resonate—they compress narrative, fear, hope, and identity into transmissible forms. They bypass logic and embed themselves in the self. From there, they grow.

To say that an idea can kill is not a metaphor. It is literal.

But how does it acquire such power?

In a sense, anatomy itself gives the game away: the body spends its finest complexity not on raw strength or filtration, but on computation. In humans, that wager has a name: the brain. No other organ has altered the course of humankind as profoundly as the human brain. Muscles can lift stone and swing hammers; lungs oxygenate blood; kidneys keep the body's chemistry within its narrow tolerances. The enigma of the human brain lies in its ability to process vast amounts of information—and to convert that information into a causal force.

We know that every thought begins as a burst of activity between neurons—a movement of ions across synapses. In Albert Einstein's brain, the foundational sentences of his Special Theory of Relativity, including the groundbreaking mass–energy equivalence formula $E_0 = mc^2$, were, at their origin, nothing more than such flickers—bioelectrical patterns, ephemeral and silent. He sat hunched over a desk, in the dim solitude of a study in early 20th-century Europe, rain tapping on window glass, smoke curling upward in silence, penning down one of the most scientifically potent ideas. No glam. No glamour. Just thought made ink. But those fleeting neural sparks did not die in the dark.

They were translated into a language. Encoded into symbols. Etched into paper. Replicated across tongues, textbooks, and decades—until they rewired the minds of millions and recharted the intellectual landscape.

This realisation forces us to confront a deeper, more unsettling question: Is information merely a byproduct of thought—or is it the true engine behind it?

Is information passive like a book on a shelf—or is it active, viral and alive, moving from neuron to neuron, from mind to mind, exerting influence like a force of nature?

Could it be that information—subtle, silent, and formless—is not merely a descriptor of reality, but a force woven into the very fabric of the universe, acting upon us with the same inevitability as tides, earthquakes, or entropy?

If so, then its fingerprints should be everywhere—etched not only in stars and circuits, but in the ways human beings think, feel, and move.

Indeed, information, in its various forms, shapes how people perceive the world and act within it—whether as political ideology, religious belief, or scientific insight. It organises thought, inspires action, and mobilises societies with a precision no less potent than the laws of physics.

The book will explore the nature of information to show that it transcends political or ideological movements to reveal a deeper, universal significance. It extends far beyond the realm of human affairs. While political systems may wield information to influence beliefs or direct power, they merely tap into a force far more fundamental. Information is not just a mechanism of communication or control—it is a foundational principle of the universe itself.

From the encoded instructions of DNA to the architecture of neural networks, from the orbits of planets to the algorithms of artificial intelligence (AI), information is the hidden scaffolding beneath form and function. It organises biology, shapes technology, and underpins the very laws of physics.

There is something quietly astonishing about the way our universe behaves. A distant star collapses and explodes, and light from that cataclysm travels across time and space to be captured by a human-made lens. A single whispered sentence can ignite revolutions, shift cultures, or alter the trajectory of a nation. Within the microscopic spiral of DNA, instructions unfold to form an eye, a wing, a tail. In all of these events—celestial, biological, social—there exists a subtler thread running beneath the surface. Beneath matter. Beneath motion. Beneath meaning itself lies something more enduring. That something is information.

We are used to thinking of information as something human-made—emails, documents, files, texts. But what if information is far older than our alphabets, deeper than our machines? What if it is as fundamental as space-time, mass or energy? What if it is not simply how we describe the world, but what the world is made of?

The central premise of this book is both radical and ancient: that information is not a passive medium, but an active principle. Across disciplines—from physics to neuroscience, from cosmology to computing—evidence is mounting that information does not merely describe the universe; it participates in its unfolding. In classical science, information was long regarded as secondary—a label we assigned to the arrangement of things. But modern developments challenge that view.

What if everything you've ever believed to be fundamental—matter, energy, space, time—is not the only foundation of reality, but there is more to it?

This book begins with a bold but increasingly supported idea: that information is more than a way to describe reality—it may be its foundation. Instead of thinking of information as something secondary, a byproduct of physical processes, we are invited to consider the opposite: that the structures we see in the world—from natural phenomenon to thoughts—may actually arise from the patterns of information underneath them.

Information is not simply a mirror reflecting what exists. It may be the invisible framework that shapes existence itself. It's not just the content of your thoughts—it may be what allows thought to happen in the first place. Much like the DNA inside a cell, information acts not only as a record but as a set of instructions: it builds, organises, and gives rise to everything we experience, from the growth of a living body to the logic behind a mathematical equation.

In quantum physics, outcomes aren't fixed in advance. Before a measurement, a photon or electron is described by a distribution of possibilities; when measured, we observe a definite result. Classic demonstrations like the double-slit show that attempts to track "which path" erase the interference we'd otherwise see. Whether this is a statement about knowledge or about reality itself depends on interpretation, but it has transformed how we think about what can be known. In this regime, information—what a measurement reveals—marks the transition from possibilities to particulars.

Modern variations, like Wheeler's delayed-choice experiment, deepen the puzzle: our choice of measurement determines which aspects of a particle's behaviour are revealed, as if the story we can tell about its past depends on how we look—without implying that the past itself is rewritten.

If so, then information becomes more than a label or a log. In that case, information is more than a label: it marks the transition from a spread of possibilities to a recorded outcome. Reality, on standard accounts, behaves like an ongoing computation in which the measurement context selects which facts are registered. The observer's apparatus doesn't merely witness; by interacting with the system, it helps fix what can be said to have happened.

Even consciousness—the apex of human complexity—emerges not as a bolt-on feature but as a late chapter in this biochemical epic. The neurons that make up our brain, the glial cells that support them, the synapses that store memory—all are built and maintained by the same informational architecture that guided the first cell division.

So, when we speak of being "made of information," it is not a poetic metaphor. It is a biological fact. Living organisms are information encoded, interpreted, and expressed. A species of sentences, shaped by syntax older than speech, speaking through flesh the language of life.

1.7 The Brain as Sculptor: Consciousness and Predictive Coding

In contemporary neuroscience and cognitive science, consciousness is increasingly theorised not as a passive byproduct of neural firing but as an emergent property of structured information processing. Rather than treating perception as a one-way channel from the senses to the brain, emerging models suggest that the brain is actively predicting, compressing, and reconstructing incoming data into coherent experiential models. This perspective is powerfully captured by the theory of predictive coding, which posits that the brain functions as a hierarchical inference engine.

According to predictive coding, the brain continuously generates internal models of the external world and uses sensory input not as raw truth but as a stream of *prediction errors*—signals that reveal the mismatch between expectation and reality. These errors flow upward through the neural hierarchy and are used to refine the brain's internal generative models, creating a loop of constant self-correction. In this framework, perception is not passively received—it is actively constructed. You do not "see" the world directly; you infer it.

The cortex, especially in the visual and sensory regions, is thus understood to be actively *predicting* what it will perceive. Drawing on prior experience and contextual cues, it generates top-down signals that anticipate incoming sensory data. When these predictions align with reality, perception proceeds smoothly; when they diverge, bottom-up signals from the senses serve to update and correct the brain's internal model. Conscious experience, in this framework, emerges at a delicate balance point—where predictions are precise enough to be efficient yet flexible enough to be revised. This dynamic equilibrium, often described in terms of information integration and precision weighting, allows the brain to construct a stable yet adaptive model of the world.

Building on this, integrated information theory (IIT) suggests that consciousness arises from the degree to which a system integrates information across its components in a unified, irreducible manner. According to IIT, it is not just the amount of information, but the way it is structured and interconnected, that matters. This reframes the "mind" not as a fixed location or object but as a *topology of informational relationships*—a network that cannot be decomposed without losing its conscious quality.

Thus, in both predictive coding and IIT, consciousness is fundamentally a *patterned structure of information.* It is less about neurons as physical units and more about the flow, integration, and compression of information within the system. In this view, the brain is less a mirror and more an artist—a self-correcting sculptor of reality, constantly revising its models based on the smallest informational discrepancies.

This epistemological shift—from matter to information, from sensory reception to predictive construction—aligns with the central argument of this book: that information is not merely a descriptor of reality but a generative force within it. If the brain constructs the world it perceives through recursive information processing, then consciousness itself becomes evidence that reality is not passively recorded, but actively shaped. In this view, the mind is not a mirror of the world, but a sculptor—its material, information.

In the technological domain, information has begun to operate independently of biological substrates. AI systems—particularly deep learning models—now ingest, classify, and generate outputs that resemble knowledge-driven tasks. But these systems do not *know* in any human sense. They possess no intentionality, no subjective awareness, no unified self to experience the content they produce. Their architecture, while conceptually inspired by simplified models of neuronal signalling, bears little resemblance to the actual structure or dynamics of the brain. These models operate as abstract mathematical systems, not biological analogues, processes data through statistical layers—matrices of weights and activations—not neurons. What emerges is not sentience but structured correlation. Language models like GPT do not understand meaning; they map probabilistic relationships between symbols within immense datasets. And yet, the outcomes—coherent sentences, plausible arguments, poetic metaphors—mimic forms of thought.

This is not imitation in the theatrical sense. It is not a simulation of human cognition in spirit or function. Rather, it is a new kind of system-level behaviour: recursive informational processing, devoid of comprehension but replete with structure. These systems do not interpret the world—they rearrange it within formal constraints. To conflate this with human-like cognition would be misleading. Yet to ignore the emergent capacity of non-biological systems to act upon information with increasing complexity would be to miss a profound shift.

In this book, *The Nature of Information*, we examine this distinction with care. Information in AI is not semantic in the way it is for minds; it does not signify across subjective experience. It is operational—systemic. Yet its consequences, its capabilities, and its imprint on society are real—and increasingly indistinguishable from the outward expressions of human cognition. The boundary between information as semantic content and information as systemic function is not vanishing; it is being reconfigured.

Philosophers have long gestured towards this deeper architecture. The ancient Greeks spoke of *Logos*—not merely as speech but as the rational order immanent in the cosmos. In the twentieth century, Gregory Bateson reframed information as "the difference that makes a difference," underscoring its generative power, not merely its descriptive use.

Today, theorists suggest that the universe is not a machine but a computation—an unfolding of information, with space and time as emergent properties.

If that is true, then what we call "reality" is just the expression of information behaving consistently. Matter becomes form. Form becomes function. Function becomes experience. Information threads through it all.

The implications are staggering. A black hole may not be defined by what it contains, but by the information it conceals. The genome is not just a string of chemicals—it is a language of potential, shaped by millions of years of evolutionary memory. Consciousness may not be a property of the brain but of any system that integrates information in a sufficiently rich and recursive way. Could intelligence—perhaps even sentience—emerge wherever information flows in self-organising patterns?

1.8 Landauer's Legacy: The Thermodynamic Weight of Knowing

Even the seemingly aloof laws of thermodynamics are not immune to the pull of information. At the microscopic level, where statistical mechanics meets logic gates, the boundary between physical law and informational structure begins to blur. This convergence is nowhere more elegantly revealed than in Landauer's principle, formulated in 1961 by physicist Rolf Landauer at IBM.

Landauer showed that erasing even a single bit of information—a binary "yes" or "no," a 1 or a 0—is not a trivial act. It carries an irreducible physical cost. To reset a memory register, to overwrite uncertainty with certainty, one must increase entropy somewhere else—by releasing a precise minimum of heat into the surrounding environment. The amount is tiny: at room temperature, about $k_B T \ln 2$ joules per bit, where k is Boltzmann's constant and T is temperature. But the implication is immense.

This is not a metaphor. It means that information is not merely an abstract construct. It is embedded in, and constrained by, the physical world. Thought, memory, and deletion—each has thermodynamic consequences.

Landauer's insight gave rise to a new understanding: computation is not divorced from physics. The act of calculating, remembering, and forgetting—these are all processes that obey the same energy rules as engines, molecules, and stars. Information, in this light, is not just a descriptor of the universe—it is a participant in its unfolding.

This revelation bridges the gap between information and thermodynamics. It tells us that the manipulation of knowledge has energetic consequences—that the mere act of forgetting, of deleting a digital file or resetting a logic gate, has an entropic cost.

In this light, entropy—the tendency of systems to move from order to disorder—is no longer just about heat and particles. It becomes intertwined with knowledge itself. When entropy increases, information is degraded or lost—structure dissolves into noise, distinctions blur. But the reverse is also true. When complexity arises, information is not just preserved—it self-organises. Molecules align into metabolic pathways. Cells specialise into organs. Neurons wire themselves into minds. And now, through human design, billions of transistors are configured into architectures that can learn, adapt, and generate output in ways once thought unique to biological intelligence.

From the spiral arms of galaxies to the spiral coils of DNA, the universe seems to favour structures that can store, process, and propagate information. Stars forge elements like carbon, nitrogen, oxygen, and phosphorus. Supernovae and neutron star mergers distribute these into the interstellar medium. Chemistry, seeded by the remnants of ancient stars, gives rise to carbon-based molecules—flexible, versatile, and stable. In the right planetary crucibles, these molecules assemble into chains: amino acids into peptides, nucleotides into strands of RNA. Some of these chains acquire the rare ability to replicate. Others fold into catalysts that assist their own reproduction. Encased in membranes of self-forming lipids, they become protocells—primitive systems where information, structure, and function begin to intertwine. Those protocells become life. Life evolves nervous systems. Nervous systems develop abstract thought. Thought invents language. Language builds civilisations. Civilisations create AI. Each stage is a leap in informational density and complexity, each one subject to the constraints and possibilities of thermodynamics.

What emerges is a picture of a cosmos that is not indifferent to meaning—but shaped by it. Information is not a passive label we attach to the world. It is a dynamic actor within it, trading energy for insight, entropy for emergence. Landauer's principle whispers this truth in every deleted file, every lost memory, and every thermal fluctuation: that the cosmos is not just a stage but a computation unfolding.

And in that unfolding, we find ourselves—systems of organised matter running recursive informational loops, fighting entropy not with brute force but with intelligence, adaptation, and memory. And so, we ask: What if the true foundation of the universe is not substance but structure? Not energy but encoding? What if, at the deepest level, everything we call "real" is simply what information does when it is allowed to flow?

In this view, you are not just a body animated by a brain. You are a temporary structure of information, woven together by billions of interactions, memories, and meanings. The mind is not a vessel—like a jug holding water—it is a process: a continuous act of sensing, integrating, predicting, and choosing. The self is an emergent identity—a continuously updated model shaped by memory, perception, and expectation.

To apprehend reality through this lens is to recognise a world in which meaning arises not in opposition to mechanism, but as its emergent expression. Where the laws of physics are not lyrical metaphors, but the algorithmic grammar of the cosmos. Where biological processes—once perceived as merely organic—reveal themselves as intricate computations of information, feedback, and adaptation. In such a view, the traditional binaries that separate machine from mind, and thought from materiality, do not collapse into mysticism; they are recontextualised through the shared language of informational dynamics. The boundaries remain—but they are rendered more permeable, more nuanced, and more cognitively honest.

This is the journey we now embark upon—not into the world of objects but into the architecture of patterns. Into the hidden grammar of the universe. Into the architecture of information itself—its internal coherences, its capacity to shape systems, and its recursive role in organising the world.

When we think of information, we often imagine it in its technological garb—numbers coursing through servers, pixels flickering on screens, lines of code compiled by programmers, or messages scrawled in ink across paper. But this conception is only a recent expression of something far older and more primal. Information existed long before humans, before language, before DNA. It existed wherever patterns were formed, sustained, and transformed.

To truly understand the nature of information, we must uncouple it from the machines it currently inhabits. Information is not just data stored in memory drives. It is not just a carrier of meaning. It is meaning's very possibility. It is what allows order to arise from chaos, memory to persist through time, and minds to emerge from matter.

In 1948, a quiet revolution took place within the pages of a Bell Labs technical journal. Claude Shannon, a brilliant yet understated mathematician and electrical engineer, published a paper titled "A Mathematical Theory of Communication." Within it, he defined information in a way that would forever alter our understanding of communication, computation, and, ultimately, reality itself. To Shannon, information was not meaning—it was a measure of uncertainty. The more unpredictable a message, the more information it contained. This deceptively simple insight gave birth to information theory, the mathematical scaffolding behind everything from the internet to AI.

At its core, Shannon's work reframed the way we think about signal and noise. Prior to him, engineers saw messages as continuous streams of electrical or radio signals. But Shannon proposed something radical: all communication could be reduced to binary digits—bits, the now-ubiquitous 1s and 0s. A bit, he defined, is the smallest possible unit of information—the answer to a yes/no question, the resolution of a single binary uncertainty.

Through this lens, a text message, a photograph, or the genome of a species are all just patterns of bits—structured resolutions of uncertainty.

Shannon's equations, particularly his entropy function, enabled engineers to calculate the optimal ways to compress and transmit data across noisy channels. If you've ever streamed a video without unbearable buffering, browsed a website halfway across the globe, or spoken clearly on a mobile phone while driving through a tunnel, you've unknowingly stood on the shoulders of Claude Shannon's work. His insights into how information behaves in the presence of noise laid the mathematical foundations for our entire digital world.

At the heart of Shannon's work lies a deceptively simple question: how can a message be transmitted with perfect fidelity through an imperfect medium? His answer came in the form of a probabilistic framework that could measure information, calculate uncertainty, and mathematically predict how much data could be compressed or how much noise could be tolerated before meaning was lost.

Modern error-correcting codes—used in everything from satellite communications to your smartphone's ability to recover dropped calls—spring directly from Shannon's limits of reliable transmission. Lossless compression algorithms like ZIP or FLAC exist because Shannon proved that redundant information can be mathematically stripped away without losing content. And encryption protocols, from secure messaging to online banking, all exploit his idea that information is quantifiable and manipulable—not just symbolic but functional.

But Shannon's true genius lay not only in engineering utility—it lay in a philosophical breakthrough. Take the example of an optical fibre. A single strand of silica glass carries pulses of light across oceans. Each flicker—on or off—is a binary digit, a bit. These light pulses are governed by the physics of internal reflection, by the speed of light, by material losses and noise. The information conveyed by the flickers of light—the structured sequence of bits—exists independently of the glass, the photons, or the electromagnetic equations. It can be counted, compressed, encrypted, and decoded, irrespective of the photons or the fibre. This conceptual breakthrough had implications far beyond engineering. It reframed the very nature of knowledge and communication. If information can exist apart from matter, then it becomes something more elemental—a force with its own grammar.

And from there emerges a deeper, almost philosophical shift: if information is, at its core, the resolution of uncertainty, then it is not just something we receive or store—it is something we produce every time we make sense of disorder. Whether in transmitting a message, interpreting a signal, or anticipating the future, information arises as the counterforce to entropy. It becomes an active agent in the structure of reality—not merely describing the world but continually reshaping it.

This principle echoes through the sciences. In biology, the genome is not merely a record of evolutionary history—it is an information archive optimised to reduce the uncertainty of survival. In neuroscience, predictive coding suggests the brain functions as a Bayesian inference engine, constantly minimising uncertainty about the world through sensory feedback.

In thermodynamics, one of the most enduring and perplexing thought experiments emerged in 1867 from the mind of James Clerk Maxwell. He imagined a tiny, intelligent being—later dubbed Maxwell's Demon—stationed between two chambers of gas. This hypothetical creature could open and close a microscopic trapdoor, allowing faster-moving (hotter) molecules to pass in one direction and slower (cooler) ones in the other. Over time, one side would heat up and the other cool down, creating a temperature gradient without doing any mechanical work. At first glance, this appeared to violate the second law of thermodynamics, which holds that entropy—a measure of disorder—must always increase in a closed system.

The paradox sparked decades of scientific debate. Was the second law inviolable, or could information somehow substitute for energy in controlling matter?

The resolution came nearly a century later, not through mechanics, but through information theory, as mentioned earlier. Physicist Rolf Landauer argued in 1961 that the demon's act of measurement, memory, and—most critically—erasure of information carries an unavoidable thermodynamic cost. Erasing a single bit of information releases a minimum amount of heat into the environment. Charles Bennett later built on this, showing that it's not the acquisition of information but its deletion that ensures the second law remains intact.

This reframed the paradox entirely: information is not inert. It has thermodynamic consequences.

What had once been seen as an abstract, immaterial concept—bits, decisions, yes–no operations—was now understood as a physical quantity, bound by the same laws that govern heat and matter. The mere act of knowing something and then forgetting it has an energetic cost.

The implications are profound. In a universe governed by entropy, information emerges not as a secondary feature but as a fundamental counterforce—a means of structuring, selecting, and resisting disorder. In

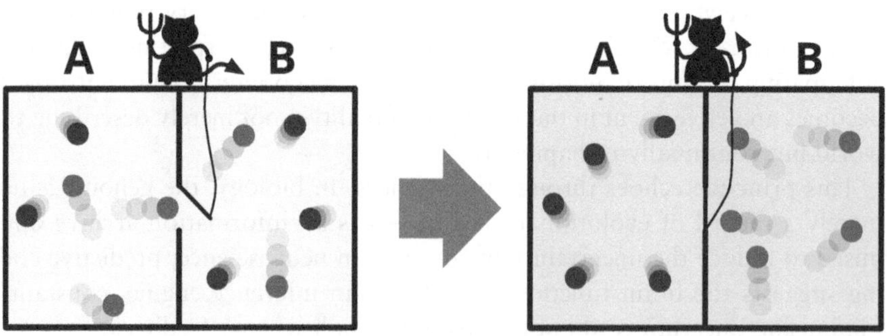

FIGURE 1.3 Maxwell's demon. A schematic showing an intelligent "gatekeeper" that opens a trapdoor to pass fast (hot) molecules one way and slow (cool) molecules the other, seemingly creating a temperature gradient without doing work and challenging the Second Law. Illustration by Htkym, via Wikimedia Commons, CC BY 2.5 (also available under CC BY-SA 3.0/GFDL).

Maxwell's paradox, we glimpse a deeper truth: that information is woven into the architecture of reality itself, not merely as knowledge but as action with thermodynamic weight.

And this is not confined to thought experiments. Where entropy seeks to erase distinctions, information draws them. It pushes back, carving out local pockets of order within the general drift towards chaos. The more complex a system becomes—whether biological, social, or technological—the more it demands structure, regularity, and coherence. And here, information steps in not just to reduce uncertainty, but to generate intelligibility.

Language, mathematics, genetic code, legal systems, and even economic frameworks can be understood as informational architectures. They do not simply store data—they compress, organise, and regulate it. They impose logic on multiplicity, taming vast fields of variables into forms that can act, adapt, and persist. In this sense, information is not just what allows life or intelligence—it is what enables complexity to hold together at all.

And once we recognise this scaffolding role, a broader panorama comes into view. Information does not merely support complex systems—it appears to anticipate them. From atoms to consciousness, from neural circuits to artificial minds, we begin to see a unifying thread: systems evolve by absorbing, refining, and compressing information. The deeper we look, the more it seems that information doesn't just run through complexity—it helps shape how complexity emerges over time. This realisation has led thinkers like John Wheeler, who famously coined the phrase "It from Bit," to suggest something radical: that information is not simply a lens through which we interpret the universe—it may be the very fabric from which the universe is woven.

If this is true, its echoes can be heard across the deepest chambers of modern science. In quantum theory, the behaviour of particles may not arise from hidden variables of position or momentum but from the informational structure of their quantum states. In cosmology, the very geometry of spacetime may be a projection of underlying informational units, as suggested by holographic principles. And in the realm of AI, machines that generate, compress, and refine information are beginning to exhibit behaviours once thought inseparable from the human mind—reasoning, creativity, and even flashes of apparent intuition.

In this light, information is no longer a passive medium passed from sender to receiver. It becomes the thread that binds cause to effect, past to future, potential to the unfolding of reality itself.

In physics, we find that the laws of the universe are describable not just in terms of particles or forces but equations—informational structures that remain consistent across space and time. The constants of nature—the speed of light, the gravitational constant, and Planck's constant—can be thought of as fixed informational values that define how the cosmos behaves.

In the quantum realm, reality doesn't behave like the predictable clockwork we observe in the macroscopic world. Instead, it dances to a subtler rhythm—a rhythm dictated not by solid particles but by waves of possibility. Here, information is not an afterthought or a descriptor. It is a fundamental player.

Quantum states are described by what physicists call probability amplitudes—mathematical functions that capture not what a particle is but what it might be. Until a measurement is made, a quantum system does not occupy a definite state. It exists in a superposition—a cloud of probabilities, a symphony of maybes. The act of measurement collapses this wave function and, in doing so, actualises a particular reality. This isn't just a philosophical abstraction—it's a measurable, repeatable feature of quantum experiments, such as the double-slit experiment, which shows that even a single electron behaves like a probability field until it is observed.

Yet nowhere is the informational nature of quantum mechanics more astonishing than in the phenomenon of entanglement. When two particles become entangled, their informational states are no longer independent. Alter the state of one, and its counterpart responds—instantaneously, regardless of the distance between them. Einstein famously dismissed it as "spooky action at a distance," but decades of rigorous experimentation have since affirmed its reality.

Crucially, this connection does not violate relativity; it transmits no matter or energy faster than light. What moves—or more accurately, what is shared—is information. A nonlocal synchrony. A kind of cosmic whisper across the fabric of space-time. Quantum information theory, a field now blossoming at the frontier of physics and computer science, seeks to understand these bizarre behaviours not as paradoxes, but as signs that information itself is the bedrock of reality. In this view, qubits—the quantum analogy of classical bits—can hold multiple states simultaneously and encode information in a fundamentally richer way. The rules of logic that govern quantum systems are no longer confined to laboratory curiosities—they are poised to reshape the future of computation, cryptography, and even our most basic assumptions about what it means to *know*.

Where classical physics offers equations of motion, quantum theory provides equations of knowledge. The shift is profound: from tracking the movement of objects to understanding the probabilities of observation; from what *is*, to what can *be known*. In this transition—from trajectory to uncertainty, from object to observer—the primacy of information becomes not just apparent, but inescapable.

Physicist John Archibald Wheeler once posed a deceptively simple question: *Could the universe be built not from things, but from answers to yes–no questions?* His answer crystallised a radical new paradigm—*It from Bit*. Reality, he proposed, does not emerge from inert substance, but from acts of informational distinction. Each quantum event is not merely a physical interaction—it is a binary decision, a moment of choice that collapses uncertainty into form. The universe is not a static stage, but a vast computation unfolding in real-time.

From this perspective, information is the invisible lattice upon which all form is suspended. It is the silent sculptor of existence, organising matter into atoms, atoms into molecules, molecules into life.

The most compelling testament to the power of information lies within us. Every living organism is a breathing, walking expression of information—encoded, replicated, and evolved over billions of years. Nowhere is this more evident than in the molecule that defines life itself: DNA.

DNA is not merely a chemical structure; it is a language. With just four chemical letters—adenine (A), thymine (T), cytosine (C), and guanine (G)—it writes the instruction manual for building and sustaining life. This manual is not stored in a book or a machine, but within the nucleus of every cell.

Each strand of DNA contains genes, and each gene is a code—a sentence within a biological programme. These codes guide the assembly of proteins, which in turn create the structures and perform the functions necessary for life. From eye colour to immune response, from growth to metabolism, all biological processes are informational in nature.

1.9 Ribosome—Nature's First Microprocessor

Any account of the universe's information fabric is incomplete without the ribosome—the cell's most remarkable nanomachine. Here, instructions are put to work and matter is shaped accordingly.

In the previous section, we considered how information, though intangible, can shape physical outcomes. Where, then, is it converted into a tangible effect? Inside the cell, the ribosome is where genetic instructions are translated into working proteins. Without ribosomes, DNA would remain a self-copying archive with no way to realise its instructions. The ribosome supplies the missing link: nature's first information microprocessor, translating compact symbols from DNA (via mRNA) into working agents—proteins that fold, catalyse, signal, and build.

A ribosome is a nanomachine that converts digital strings of nucleotides into three-dimensional working matter. It reads the four-letter alphabet of messenger RNA (mRNA) and, line by line, writes the 20-letter language of proteins by stitching amino acids into precise sequences (Crick 1958). In doing so, it executes the central translation in biology—the only route from genetic information to the catalytic, structural, and regulatory molecules that run cells. No ribosome, no proteome; without it, genomes are silent manuscripts.

Ribosomes are universal (Ban et al. 2000; Schmeing and Ramakrishnan 2009; Yusupov et al. 2001). Bacteria, archaea, and eukaryotes all carry them; mitochondria and chloroplasts house streamlined versions that echo their bacterial ancestry. This universality is not merely a quirk of shared descent—it reflects the fact that biochemistry has converged on a single solution to the problem of mapping codons to amino acids with high speed and fidelity in water. The ribosome's centrality is reflected in cellular economics: in rapidly growing bacteria, translation can consume the majority of the cell's energy budget, and ribosomes can account for a large fraction of cellular mass. Polysomes—trains of ribosomes translating the same mRNA simultaneously—illustrate the scale and coordination: the cell's information factory floor (Ingolia 2011).

Yet the ribosome is more than a fabricator; it also serves as an information microprocessor. Its cycle would look familiar to a computer architect: fetch (advance to the next codon), decode (select the matching tRNA), execute (form the peptide bond), and write-back (shift forward and update the state for the next instruction). It uses dedicated "coprocessors" (translation factors) for timing and error control, layers multiple proofreading steps to keep errors near 10^{-4}–10^{-3} per codon (Rodnina 2018; Zaher and Green 2009), and—when many ribosomes translate the same message—runs in parallel as polysomes. Its instruction set—the genetic code—is nearly universal with only minor dialects. The ribosome is thus the first and most consequential information microprocessor that nature built.

Seen up close, think of the ribosome as two master craftsmen in adjacent rooms, working in lockstep with distinct roles. One craftsman is the small subunit, the expert reader of the message; the other is the large subunit, the builder that links amino acids into a chain. In bacteria these are called 30S (reader) and 50S (builder); in eukaryotes, 40S and 60S. The numbers are Svedberg units—how quickly each part sediments in a centrifuge—and they reflect shape as well as mass. Despite the different labels, the architectural plan is conserved across life.

Each subunit is built around a core of ribosomal RNA (rRNA). These long RNAs fold into intricate, layered scaffolds. Ribosomal proteins—dozens of them—wrap the rRNA like braces and hand-holds: they neutralise charges, clamp helices in place, and tune the motion between parts. The proteins make the machine sturdy and controllable; the RNA provides the frame and, at the centre, the chemistry.

Across the seam where the two subunits meet run three stations for tRNAs—A, P, and E. The A (aminoacyl) site is the arrival gate for the next amino acid; the P (peptidyl) site holds the growing chain; the E (exit) site is where the empty tRNA leaves. As the message advances, tRNAs step A → P → E at a measured pace, like workers handing a part down a line.

Although the ribosome carries many proteins, the active heart is RNA. High-resolution structures show no protein side chain in the peptidyl transferase active site (Ban et al. 2000; Nissen et al. 2000). The ribosome's signature reaction—forming the peptide bond—therefore belongs to rRNA itself. It is, in the strict sense, a ribozyme.

The catalytic centre of the ribosome is rRNA, which mediates peptide-bond formation. Accurate protein synthesis, however, first requires correct start-site selection. During initiation, the small subunit engages the mRNA and identifies start signals that position the initiator tRNA in the P site and align the reading frame. Once this register is established, the large subunit joins, and elongation proceeds codon by codon.

The ribosome must choose the right place to begin. In bacteria, a short tag just before the first word of the message helps the reader line up precisely (biologists discovered that this tag pairs with a matching patch on the ribosome's reader RNA) (Hinnebusch 2014). In eukaryotes, the reader latches onto the front end of the message and slides forward until it finds a good "start" word. In both cases, the special starter tRNA is placed in the P site, the A site is empty, and the machine is primed. (Bacteria can sometimes start at near-look-alike start words, but less efficiently.)

Every three-letter word (codon) triggers a four-beat routine (Hopfield 1974; Ninio 1975; Rodnina 2018):

i. **Bring.** A delivery helper carries in the matching tRNA loaded with its amino acid (called EF-Tu in bacteria, eEF1A in eukaryotes). The anticodon tests the codon; close but wrong matches fall away.
ii. **Check.** If the match is right, the ribosome spends a bit of energy to double-check before committing—an error-cutting trick known as kinetic proofreading (Hopfield 1974; Ninio 1975; Rodnina 2018; Zaher and Green 2009).
iii. **Join.** In the builder subunit's active site, the growing chain is transferred onto the new amino acid. Remarkably, the catalyst here is RNA itself (Rodnina 2011; Schmeing et al. 2005).
iv. **Move.** A second helper (**EF-G** in bacteria, **eEF2** in eukaryotes) uses energy to shift the ribosome one codon along the message; the tRNAs step to their next seats (Voorhees and Ramakrishnan 2013).
v. **Finish and reset.** A stop signal in the A site brings in a release helper—one kind in eukaryotes and two related kinds in bacteria—which snips the new protein free (Dever and Green 2012; Schmeing and Ramakrishnan 2009). Finally, "splitter" helpers open the machine so the two halves separate; they can then reassemble at a new message and begin again.

Translation must be both accurate and fast. Each amino acid that is added costs energy: two GTP per elongation cycle—one to deliver the tRNA and one to move the ribosome forward—and, upstream, one ATP is split to AMP when a tRNA is charged by an aminoacyl-tRNA synthetase (Ibba and Söll 2000). Typical speeds are well measured: bacteria add about 10–20 amino acids per second at 37°C, while yeast and mammals add ~3–8 per second (Ingolia 2011; Rodnina 2018).

Accuracy comes from checks at several stages. First, aminoacyl-tRNA synthetases edit misactivations or mischarged tRNAs. Next, the ribosome verifies codon–anticodon pairing and uses kinetic proofreading to reject near-matches after GTP hydrolysis (Hopfield 1974; Ninio 1975; Rodnina 2018; Zaher and Green 2009). Finally, surveillance systems remove faulty messages or clear stalled ribosomes (Shoemaker and Green 2012). In short, cells spend energy to buy accuracy, and layered checkpoints turn a noisy chemical process into reliable protein-making (Rodnina 2018).

The ribosome is such an awe-inspiring structure that it compels the question: How did nature assemble such a molecular foundry? Research shows that the ribosome in its present form is the result of billions of years of evolution. The structure was created in parts—elaborated in the following text.

How did nature sculpt a nanomachine of such exquisite precision? Not by a single leap, but incrementally, through accretion over aeons. Over billions of years, selection kept what worked and layered new parts onto an older core. At the centre sits an RNA catalyst that could already stitch amino acids. Around it, evolution gathered adapters that delivered the right building blocks, a reader that enforced pairing rules, timing modules that drove the cycle forward, and protein scaffolds that stabilised and accelerated the whole.

What follows is a phased reconstruction—drawn from structure and biochemistry—of how a small catalytic kernel could accrete, piece by piece, into the programmable ribosome we know today.

Structural biology delivered the crucial clue: the modern peptidyl transferase centre (PTC) is made of RNA (Ban et al. 2000; Nissen et al. 2000; Petrov et al. 2014). Early crystallographic and cryo-EM studies showed that the chemistry of peptide-bond formation occurs in an rRNA-dominated cavity; proteins are excluded from the active site. This observation aligns with the RNA world framework and with laboratory selections demonstrating that ribozymes can aminoacylate tRNAs or tRNA mimics and catalyse peptide-bond formation in simplified contexts (Saito, Kourouklis, and Suga 2001; Turk, Chumachenko, and Yarus 2010). In other words, RNA alone can, in principle, carry out translation-like chemistry.

i. **In Phase I:** A small catalytic RNA (ancestral PTC) could have promoted condensation between activated amino acids attached to short RNA adapters—proto-tRNAs. The "decoding" in this epoch was likely crude: stereochemical affinities between certain amino acids and RNA triplets, plus the geometry of mini-helices, could bias pairings. Catalysis would have been slow and error-prone but locally advantageous—producing short peptides that stabilise RNAs and membranes, feed back into catalysis, and increase protocell fitness.
ii. **Phase II:** As peptide length increased, selection favoured a consistent mapping between triplets and amino acids. Proto-tRNAs matured

into L-shaped adapters with acceptor stems for charging and anticodon loops for pairing; their uniform geometry let a single ribozyme architecture handle diverse substrates. Aminoacyl-tRNA synthetases—today protein enzymes—may have been preceded by ribozymes or ribonucleoprotein catalysts in an earlier RNA-world stage, before specialized protein 'compilers' evolved to charge adapters with the correct amino acids. The genetic code likely crystallised through the interplay of three forces (Crick 1968; Knight, Freeland, and Landweber 1999; Wong 2005): (i) stereochemical tendencies (certain codons/anticodons weakly liking certain amino acids), (ii) biosynthetic coevolution (pathway precursors and related amino acids mapping to related codons), and (iii) error minimisation (arrangements that lessen the phenotypic impact of point mutations and misreads). The result is a code remarkably robust to noise.

iii. **Phase III: Subunit differentiation, GTPase timing, and processivity**
With a reliable adapter system in place, efficiency became king. The catalytic RNA scaffold expanded and differentiated into functional neighbourhoods: a decoding centre for geometric scrutiny of base pairs and an active site for chemistry, bridged by corridors for tRNA movement. Two subunits diverged to specialise: the small subunit optimised for RNA–RNA recognition and reading, the large subunit optimised for catalysis and exit of the nascent peptide (Petrov et al. 2014; Yusupov et al. 2001). Dedicated GTPases (EF-Tu/eEF1A and EF-G/eEF2) joined as timing devices that couple nucleotide hydrolysis to conformational changes, creating a ratcheting cycle that enforces directionality and boosts throughput while maintaining accuracy (Voorhees and Ramakrishnan 2013).

iv. **Phase IV: Protein accretion and specialisation; eukaryotic expansion segments**
Proteins progressively infiltrated the rRNA scaffold not to run the core chemistry but to solve engineering problems: stabilise flexible rRNA helices, neutralise charges, tune dynamics, and interface with cofactors and membranes. The ribosomal protein inventory diversified; basic extensions snake into rRNA grooves like molecular rebar. In eukaryotes, the core was overlaid with expansion segments and additional proteins that integrate translation with mRNA processing, surveillance, and organelle targeting, while mitochondria and chloroplasts retained bacterial-like cores adapted to their bioenergetic niches.

v. **Phase V: Integration with cellular information economy**
Finally, translation is integrated with transcription, RNA processing, and degradation to create a continuous information pipeline. In bacteria, transcription–translation coupling allows ribosomes to load onto mRNA as it emerges from RNA polymerase, linking gene expression directly to protein output. In eukaryotes, compartmentalisation (nucleus–cytoplasm) introduces checkpoints—capping, splicing, polyadenylation, export—that gate which messages get processed on ribosomes. Cells

learned to steer ribosomes to the right membranes (ER) at the right times (co-translational translocation), to pause strategically to allow folding, and to coordinate production with demand via signalling pathways (mTOR, GCN2) (Proud 2019; Sonenberg and Hinnebusch 2009).

Taken together, the ribosome we see today is a palimpsest. At its core is an RNA enzyme that likely predates DNA genomes and protein enzymes; around that ancient kernel, evolution layered timing circuits, structural supports, and interfaces that transformed a primitive peptidyl transferase into a fast, accurate, programmable microprocessor for molecular manufacturing.

The computing language here is not decoration; it is a precise mapping between informational primitives (symbols, registers, clock, instruction cycle) and biochemical realisations (codons, tRNAs, GTPase cycles, elongation). This bridge lets the chapter generalise beyond biology when the book later compares ribosomes to other natural interpreters.

mRNA is a one-dimensional instruction tape. It carries opcodes (codons) and addresses (start/stop signals, SD or Kozak context). Codon degeneracy and biased codon usage act as error-mitigating encodings that tune speed and fidelity; synonymous codons are not equivalent at runtime because they modulate elongation kinetics and co-translational folding. Untranslated regions supply headers and footers that control queueing, localisation, and lifetime of the instruction.

tRNAs are the registers: uniform L-shaped adapters that load (are charged with) specific amino acids by aminoacyl-tRNA synthetases (the compiler that enforces type safety). The anticodon loop is the address line; the acceptor stem is the data payload. The ribosome's corridor between A, P, and E sites is the bus that moves registers through the execution unit.

The PTC is the arithmetic–logic unit of biochemistry. It performs one operation—condensation to make peptide bonds—but with high efficiency and in a controlled, discretised cycle. EF-Tu/eEF1A and EF-G/eEF2 serve as a clock: GTP binding/hydrolysis provides ticks that gate state transitions (accept → accommodate → catalyse → translocate). Inter-subunit rotation and head swivelling are the mechanical oscillators that couple chemical energy to directed motion, ensuring that the machine advances exactly one codon per tick.

There are branch instructions: start sites, programmed translational frameshifts, readthrough of stop codons, and internal ribosome entry sites that redirect initiation logic (Atkins and Gesteland 2010; Caliskan et al. 2014). There are interrupt handlers: rescue pathways that free stalled machines and dispose of corrupt outputs. There is speculative execution of a sort in kinetic proofreading: the system commits resources to a candidate aa-tRNA and then rolls back if geometry or timing is off, paying an energy tax to reduce errors.

Ribosome pipelines work by running many copies in parallel (polysomes), with nascent chains emerging like products from staggered stations on an assembly line. Cells modulate throughput by changing ribosome number (biogenesis), elongation rates (factor availability, tRNA pools, codon usage), and message lifetime (decapping, deadenylation) (Kressler, Hurt, and Bassler 2010; Shajani, Sykes, and Williamson 2011). Feedback loops maintain balance with metabolism and stress responses, just as operating systems throttle processes to keep a computer responsive.

Nature uses several layers of error checking: (i) editing pockets in aaRS that hydrolyse misactivated amino acids or mischarged tRNAs, (ii) geometric scrutiny in the decoding centre with induced fit, (iii) kinetic proofreading after GTP hydrolysis, (iv) termination-only at bona fide stop codons enforced by release factors, and (v) after-the-fact quality control of aberrant messages and nascent chains. The net effect is a code-execution engine that produces reliable proteins in a noisy, thermally agitated medium.

The ribosome's economy of design shows how minimal informational primitives—a standardised symbol set, one catalytic operation, a clock, and error checks—can generate maximal functional diversity. That is the book's core claim in miniature: information architectures, once established, scale into worlds of possibility.

What makes the ribosome beautiful is not just its size or antiquity, but the economy of its design. With RNA as the catalytic core, it achieves chemistry without side chains at the active site by using geometry, electrostatics, and water to do the heavy lifting. With tRNAs as standardised adapters, it multiplexes one catalytic centre across all 20 substrates. With GTPases as timers, it converts chemical-free energy into sequential logic. With sequence features in mRNAs and nascent peptides, it links computation to context. And across billions of years, it has remained conservative at the core while evolving flexible peripheries that tune performance for bacteria in hot springs, neurons in the cortex, and chloroplasts in leaves.

This is what a microprocessor built by chemistry looks like: a machine that turns strings into function, disciplined by error control, clocked by GTP hydrolysis, and embedded in a larger operating system of gene expression and cellular physiology. Having outlined information execution at the molecular level, we now turn to information control across cells and time: how DNA programmes embryogenesis.

1.10 DNA and Embryogenesis: Information Writing Flesh

Embryogenesis is not merely the unfolding of a genetic script—it is the performance of a biological symphony, conducted by information itself. From the moment fertilisation starts, a single cell is set on a path of extraordinary transformation. And yet, the genome it carries is not a rigid blueprint—it is a dynamic system, responsive, conditional, and exquisitely sensitive to context.

Within hours, the zygote begins to divide, but those divisions are not repetitions—they are decisions. Genes are not simply expressed; they are regulated, activated, silenced, and modulated in response to molecular signals, positional gradients, and feedback from neighbouring cells. As lineages diverge, some cells begin to fold inward; others migrate across the embryonic axis, guided not by instinct but by encoded protocols—cascading networks of regulation that determine not just what a cell will become but when and where it will become it.

This is cellular differentiation: a precise, recursive, and self-adjusting choreography of information. The same genome—identical in every nucleus—yields neurons, muscle fibres, blood, and bone, not by altering its content but by interpreting its content differently depending on its informational environment.

What emerges is not the deterministic execution of a static code but a fluid negotiation between instruction and interaction. Life begins not just with DNA but with the architecture of meaning that DNA can generate when entangled with time, space, and signal. In this light, embryogenesis becomes a profound statement: that complexity does not arise from an abundance of matter but from the refinement and contextualisation of information.

But this orchestration does not end at birth. The informational processes that shape a body in the womb continue across generations. Evolution, in this view, is not merely the survival of the fittest—it is the continuous editing of informational code. Through mutation, recombination, and the sieve of natural selection, DNA is revised and optimised to meet the demands of ever-shifting environments. Life doesn't simply endure—it experiments, adjusts, and remembers. Species, over time, learn—not consciously but through encoded consequence—how to better inscribe the logic of their survival.

Just as computer programmes evolve through patches and updates, genomes evolve through informational improvisation. Life, in this view, is not a miracle of matter. It is a masterpiece of information management.

As life advanced, information found new ways to travel. The earliest organisms communicated chemically—bacteria releasing molecules into the environment to coordinate behaviour. Over millions of years, life developed nerves, hormones, and senses. In time, consciousness emerged and, with it, the most complex information channel ever known—language.

1.11 Language and the Hunt: Evolution's Entropy-Reducing Leap

Language was a profound evolutionary leap. It allowed abstract ideas to be shared, memories to be preserved beyond the lifespan of a single brain, and experience to ripple outward across generations. With speech came storytelling. With writing, permanence. With printing, revolution. Each new medium of expression was not merely a tool—it was a transformation of how information flowed, and what it could become. At its core, language is an entropy-reducing system. It takes the vast, ineffable ambiguity of thought and compresses it into structured sequences—symbols, sounds, grammar. Each word narrows a field of meanings; each sentence transmits a selective sliver of inner life. In this sense, language mirrors DNA: both encode complex, adaptive knowledge into a replicable, low-entropy format. Both are engines of continuity. Both are scripts for survival.

But to understand why language emerged in the first place, we must return to a more primal arena: the hunt.

Humans, unlike other predators, are not particularly fast or powerful. We lack sharp claws, prominent canine teeth, or thick protective hides. We are neither as swift as gazelles nor as strong as gorillas. And yet, *Homo sapiens* rose to become the planet's apex predator. The secret was not in muscle—it was in cooperation. And cooperation, in turn, required communication.

As social primates, early humans faced a collective action problem: how to coordinate group behaviour towards shared goals, especially in high-stakes activities like hunting. Other pack predators—wolves, hyenas, lions—rely on visual signals, vocalisations, and body language to coordinate their attacks, synchronising movements with remarkable precision but limited preplanning. But these methods, while effective, are limited to the here and now. Human language added something radically new: the capacity to plan future actions, allocate roles, and refine strategy before the moment of confrontation.

This idea is supported by the work of Michael Tomasello, who argued that shared intentionality—the ability to represent common goals in the minds of multiple individuals—was a key selective pressure in the evolution of language and cognition (Tomasello 2008). Similarly, Robin Dunbar's social brain hypothesis proposes that language emerged as a social grooming tool, eventually expanding to manage group cohesion at larger scales (Dunbar 1996). But more fundamentally, language reduced entropy in the collective effort—transforming a chaotic scramble into a choreographed act.

The hunt was where the physics of thermodynamics met the biology of cooperation. In predatory events, time and energy are everything. A failed chase costs calories, exposes the group to danger, and erodes social trust. The "rabbit chase problem," observed in predator–prey dynamics, illustrates how prey often escape not by strength but by drawing out the chase until the predator collapses from exhaustion. In the Pleistocene landscapes where Homo sapiens evolved, survival often hinged on the success of the hunt. A misstep, a mistimed flank, or a missed cue could mean not just failure—but hunger, vulnerability, even death. Language, then, was not just a cognitive tool—it was an energy-saving mechanism. Language, in this context, was not a luxury. It was a lifeline. A low-entropy protocol to increase the signal-to-noise ratio of group behaviour. Evidence increasingly suggests that complex language emerged during the hunter-gatherer phase of human history, when coordination, tool use, and social cohesion became evolutionary imperatives. Archaeological findings link advancements in vocal control and symbolic thinking with the demands of cooperative foraging and hunting. For instance, research by Stout et al. (2008) on the neural underpinnings of stone toolmaking shows a strong overlap between motor planning and language circuits in the brain, indicating that the development of complex tool use may have co-evolved with linguistic capability. Similarly, philosopher of biology Kim Sterelny (2012) argues that early human societies depended heavily on shared intentions and collaborative behaviours—pressures that likely selected for increasingly sophisticated means of symbolic communication.

In such environments, language offered a unique evolutionary edge. It enabled pre-hunt planning, role assignment, and strategic foresight—functions that nonverbal cues alone could not support. Unlike wolves or lions, whose coordination depends on real-time signals like body posture or gaze, humans could model the future. They could discuss terrain, anticipate prey behaviour, distribute risk, and refine tactics based on past failures. Each utterance reduced uncertainty. Each sentence synchronised minds across space and time. In effect, language became humanity's first internet—an invisible network that linked separate brains into a single, coordinated intelligence.

In this way, language dramatically increased the efficiency of group behaviour. It minimised wasted motion, duplicated effort, and fatal confusion. By streamlining group cooperation, it reduced the effort required and increased the payoff of working together. From a systems perspective, it stabilised cooperation by making intentions legible—reducing wasted effort, aligning expectations, and turning collective action into reliable success. Long before written records, language was shaping the flow of information through human networks—compressing complexity into coordination, and noise into survival.

In this light, the earliest symbolic expressions take on new significance. The cave paintings at Chauvet (France), Lascaux (France), and Bhimbetka (India)—some over 30,000 years old—are not idle art. They are records of memory, risk, and method. Etched with ochre and charcoal, they document not only what humans hunted but how they did it. These were informational blueprints, not unlike software: compressed, symbolic models of reality meant to be interpreted and reenacted.

Such expressions suggest that the first use of symbolic thought—whether verbal or visual—was not metaphysical but tactical. It enabled humans to externalise strategy, store experiential data, and optimise future outcomes. In other words, it allowed information to become infrastructure.

Thus, like thermodynamics, language emerged to curate order from chaos, not in atoms but in action. It became an invisible scaffold for trust, alignment, and shared risk. A means by which complexity was made manageable—by modulating the flow of informational entropy across a group. In this context, language was not just about meaning—it was about metabolism. It organised bodies into behaviours. It structured how energy was spent and how danger was avoided.

And from those early murmurs—whispers in the dark, instructions at a kill site, from those early stories told around the fire came the scaffolding of civilisation—villages, laws, trade, mathematics, religion. Each was not merely a cultural artefact but a new format for encoding and transmitting information. Each layer deepened a thermodynamic impulse: to reduce uncertainty, align human behaviour, and structure survival against the randomness of nature.

Seen through this lens, the emergence of language was not a miracle. It was a phase transition—a predictable leap in the physics of complex life. Whenever organisms must coordinate under constraints, something akin to language inevitably emerges to tame entropy and enable cohesion.

1.12 Flow Funds Form: Energy, Information, and the Eventful World

Think of reality as a three-thread braid: energy, information, and events. Energy is the capacity to do work—the shove that moves systems away from stasis. Information is patterned difference—a structure that can be read, stored, and acted on. Events are state changes in time—the ticks by which any system rewrites itself. Where these threads align, complexity blooms; where they slacken, worlds go quiet.

This sounds poetic until you watch what happens when flows rise or fall. A river fed by meltwater braids into stable channels; close the source and the channels silt up and choke. A brain richly perfused with oxygen learns; starve it and patterns unravel. A city with steady power and clean water keeps laws, schedules, and networks in sync; pull the plug and everything descends into mayhem. None of this is mystical. It is the ordinary physics of far-from-equilibrium systems: if you keep useful energy flowing through (not just stored in) a structure and give it channels for feedback and repair, the structure can maintain information—the rules, memories, and constraints that make it what it is (Nicolis and Prigogine 1977; Prigogine 1977).

i. **Energy drives events**
   Every event you can name—an electron jumping between levels, a chemical bond breaking, a protein folding, a neuron firing, a tectonic plate slipping—requires enough energy to cross a barrier. In chemistry the barrier is activation energy; in electronics it is a threshold voltage; in muscle it's the threshold membrane potential that triggers an action potential. Increase throughput—usable energy per unit time—and more barriers can be crossed per second. If you could attach a counter to things and

click it at every physical or chemical change, a river stone would tick rarely; an earthworm would chatter; a beehive would hum.

This is the first link in the braid: more flow → more events. With no incoming energy and no gradients to spend, counters stop. In thermodynamic language, the system relaxes towards equilibrium, where nothing persistent happens because nothing has the budget to happen.

Equilibrium is tea that goes cold; far-from-equilibrium is the kettle that keeps boiling because the burner continues to deliver heat. The second regime is where structure appears.

ii. **Events carry information**

A system's "event stream" is how it writes and updates itself. In a cell, events are ion channels opening, enzymes clicking substrate to product and small bursts of gene transcription. In the brain, events are spikes and pauses, synapses strengthened or weakened by a whisker. In weather, events are the births and deaths of vortices. In societies, events are messages sent, purchases cleared, laws passed, votes cast, doors opened and closed on time.

When events in a system are independent, what you see is randomness—like static. But when events begin to link up—this one triggers that one, that one suppresses another—patterns appear. Cause and effect starts to organise the chaos. The physicist Hermann Haken called this synergetics: keep energy flowing through a system and a few high-level "control knobs" emerge, steering the swarm so countless tiny parts start moving together in simpler, coordinated ways (Haken 1977). Steven Strogatz popularised a close cousin in synchronisation: fireflies blinking together, pacemaker cells beating in unison, power-grid oscillators locking phase (Strogatz 2003). Each case is a small miracle of economy. A trickle of energy keeps signals flowing, and the signals keep the events in step.

Seen this way, information is not a ghostly add-on. Information is regularity in the event stream—statistical dependencies you can bet on. Shannon's definition—information reduces uncertainty—fits perfectly: if after X, Y is likely, you have captured information (Shannon 1948). If you can store that regularity—DNA, synaptic weights, a legal code—and reuse it later, you have memory. Memory is an informational structure that outlives the event that created it.

iii. **Information shapes future energy use**

Now the braid closes. Stored regularities bias which events are likely next, steering energy into particular channels. A gene network up-regulates one pathway and down-regulates another; a neural circuit routes glucose into spikes that carry a learned solution; a traffic code funnels cars into flows that minimise collision and maximise throughput; a platform policy routes attention into some clusters and not others. Patterns become causes.

> Philip Anderson's slogan—"more is different"—was pointing at this leverage (Anderson 1972). When you stack events into persistent patterns, new levels of description appear with their own laws. Robert Laughlin called these protectorates: higher-level regularities that are robust to the microscopic chaos underneath (Laughlin and Pines 2000). You do not need every molecule of air to compute a hurricane's eyewall; you need the right variables—pressure, temperature, angular momentum—and the continuing flow that keeps the structure in being.

This three-link logic—flow → events → information → flow guided—is the skeleton we now put muscle on.

The classroom classic is the Bénard cell. Heat a shallow layer of fluid from below. At first you see random shimmering. Turn the heat past a threshold, and hexagonal convection rolls appear, stable and beautiful: warm plumes rise, cool plumes sink; heat is transported efficiently; the pattern self-selects because it is a good conductor for that flow (Nicolis and Prigogine 1977; Prigogine 1977). Turn the heat off, and the pattern melts. Turn it up too far, and neat cells give way to turbulence. Order exists in a window where flow is high enough to support structure but not so high as to shred it.

Scale up, and you hear the same tune. Hurricanes and jet streams are dissipative structures—coherent patterns that exist because they are effective routes for moving gradients towards equality (hot to cold). Rivers self-organise into braided channels; flame fronts form coherent reaction zones; dunes arrange themselves by wind. None is alive. All are informational in the physical sense: regular, rule-governed ways of conducting energy through matter.

Life takes the same principle and layers memory on top. Schneider and Sagan put it bluntly: living systems are expert exporters of entropy (Schneider and Sagan 2005). A leaf takes in low-entropy sunlight, routes it through photosynthesis to build sugars, and dumps heat. A bacterium burns glucose to maintain proton gradients across its membranes, using those gradients to drive ATP synthase, which in turn fuels repair. A human liver does endless housekeeping: recycling molecules, detoxifying, rebuilding, flushing heat.

Repair is central. Structures decay. Proteins misfold. DNA suffers lesions. Membranes leak. To keep information accurate and usable, organisms must measure deviations and push back. That pushback—error detection and correction—has a thermodynamic price. Landauer made the cleanest bridge: erasing a bit—resetting a register to a standard state—has a minimum heat cost ($k_B T \ln 2$) at temperature T (Landauer 1961). Bennett showed how reversible computation can, in principle, avoid that cost until information is erased (Bennett 1973). Real devices and living cells pay much higher costs, but the moral is clear: information processing is embodied, and memory upkeep is never free. Even in brains, much of the energy budget pays not for "thinking hard" but for keeping ready: resetting ion gradients, maintaining resting potentials, replacing turned-over proteins (Attwell and Laughlin 2001; Raichle and Gusnard 2002).

From here Chaisson's notion of energy-rate density—Φ, usable power per unit mass—offers a helpful, if coarse, cross-scale lens on how much "doing" a kilogram of system can sustain (Chaisson 2001, 2011). Think of Φ as the *throughput budget* available to keep patterns accurate, to sense and decide, to repair and reorganise. The claim is not that more energy automatically yields more order; it's that sustained, well-coupled throughput makes it *possible* to maintain richer informational organisation. "Well-coupled" matters: the flow must actually be harnessed by feedback and memory, not simply pass through as heat.

What does "coupled" look like in practice? In any complex system, energy must be metered into control loops. Those loops perform three basic jobs: (i) sensing (measuring deviations from a target), (ii) computing (deciding a corrective action), and (iii) actuating (doing the work to push back). Each job consumes power. With a larger, steadier Φ, a system can afford more sensors sampling more frequently, finer-grained models that track more variables, and actuators that correct faster and more precisely. In information terms, higher Φ buys lower noise, fresher data, tighter tolerances, and more layers of memory (short-term buffers, mid-term caches, long-term archives) to guide future action.

Seen this way, Chaisson's cross-scale comparisons are intuitive. Stars and galaxies exhibit moderate Φ: lots of power, but spread over enormous mass, with relatively simple "controllers" (gravity, radiation hydrodynamics). Active planets step up in Φ by channelling stellar flux through atmospheres and oceans—jet streams, convective cells, chemical cycles—adding feedbacks that stabilise climate. Cells route chemical energy through membranes and enzymes; their Φ is higher because they run dense control stacks (gene regulation, metabolic checkpoints, error-correcting replication). Brains sit higher still: per kilogram, they continuously sense, predict, and correct. Cities and digital networks often top the chart: extraordinary power per kilogram routed through deeply layered control architectures—markets, laws, standards, software—which keep coordination accurate in noisy environments (Chaisson 2001, 2011).

This lens clarifies a familiar puzzle: why brains are hot. The human brain is ~2% of body mass yet consumes ~20% of resting metabolic energy (Attwell and Laughlin 2001). Most of that power is not spent on dramatic "Eureka!" moments but on housekeeping for precision. Neurons must continually reset ion gradients across their membranes with ATP-hungry Na^+/K^+ pumps so that spikes are cleanly separated from silence. Synaptic transmission adds its own bill: presynaptic vesicle cycling, neurotransmitter synthesis and reuptake, postsynaptic receptor currents, and the signalling cascades that adjust synaptic strength. Even in quiet states, networks pay a standing charge to maintain readiness—resting potentials, baseline transmitter levels, glial support, and the ongoing replacement of worn proteins (Attwell and Laughlin 2001; Raichle and Gusnard 2002). In short, to keep a high-fidelity code running in a warm, wet, noisy substrate, the brain must spend continuously.

Where, exactly, does that energy go? A simplified breakdown helps. First, ionic homeostasis: each spike dumps sodium into the neuron; restoring the resting distribution costs ATP via the Na^+/K^+-ATPase. Second, synaptic work: loading vesicles, docking and fusing them, clearing transmitter from the cleft (via transporters or enzymes), and re-acidifying vesicles all draw on cellular power. Third, plasticity: learning requires protein synthesis, receptor trafficking, cytoskeletal remodelling—materials and transport costs that follow activity. Fourth, support systems: astrocytes and other glia shuttle lactate, buffer ions, and recycle neurotransmitters; microglia prune synapses; endothelial cells maintain the blood–brain barrier. Each service is part of the control stack that keeps neural signals interpretable (Attwell and Laughlin 2001).

The same thermodynamic story scales up to explain why cities are hungry. A functioning metropolis continually exports entropy—in waste heat, effluent, wear—while preserving the accuracy of its internal codes: addresses, maps, schedules, legal records, inventory ledgers, account balances. That preservation is physical. It requires steady energy for pumping and purification (water, sewage), temperature control (refrigeration, HVAC, data-centre cooling), transport (fuel for logistics and public transit), communication (cell towers, fibre backbones, switching equipment), and institutional control loops (policing, courts, regulatory reporting). The "rulebook" of a city—laws, standards, protocols, software—is written into silicon, paper, steel, and concrete. Keeping it coherent is ongoing work: filing and auditing records, recalibrating sensors and meters, patching software, resurfacing roads, and inspecting bridges. Each task is an error-correction cycle; all of them together are the city's informational metabolism.

Crucially, high Φ is enabling, not guaranteeing. Without good coupling, extra power becomes turbulence, congestion, or waste heat. A blackout doesn't merely darken buildings; it deletes coordination as traffic signals fail, pumps stall, databases desynchronise, and the city's memory fragments. Conversely, modest increases in well-coupled flow can unlock disproportionate gains: tighter traffic signal timing cuts idling and collisions; reliable water pressure stabilises hospital operations; cleaner baseline power lowers data-centre cooling loads and error rates. In both brains and cities, throughput funds upkeep, and upkeep preserves the patterns that make future throughput useful.

There are limits and windows. The brain's precision degrades under fever or hypoxia because molecular tolerances are exceeded—pumps slip, receptors misgate, and noise rises. Something analogous happens in cities during heat waves or unregulated surges in demand: transformers trip, rails buckle, servers throttle, and emergency lines clog. The cure is not simply more power; it is designed coupling—buffers, redundancy, throttles, and bright-line protections—so that added energy enters the right channels. That is exactly the point Chaisson's Φ helps you see: higher energy-rate density *enables* richer information, provided the system pays the control costs to measure, decide, and repair.

Finally, return to the central thesis: information is constraint—stable patterns that bias what happens next. Upholding those patterns requires continual work and energy. Maintaining constraint is expensive. Landauer's principle gives a lower bound for erasing a bit in an ideal device, but real systems—from synapses to court archives—operate far above that bound; they must over-sample, cross-check, and rehearse to stay accurate (Landauer 1961; Bennett 1973; Attwell and Laughlin 2001). Brains "spend to be ready"; cities "spend to stay synchronised." In both, sustained, well-coupled flow pays the continuous bill for memory with consequences—the only kind of information that can steer tomorrow's events rather than merely describe yesterdays.

Nowhere is this budget logic more concrete than in the nervous system. A single neuron is noisy: its spikes are probabilistic, shaped by thousands of inputs and its own ion-channel dynamics. Its membrane potential jitters with thermal noise; thousands of synapses bombard it with inputs of varying strength and timing; the same stimulus can make it fire on one trial and miss on the next. What brains do, reliably, emerges when many such noisy units are pooled. Two simple tricks do most of the work. First, averaging: when many neurons with similar preferences (say, for the same edge in the visual field) project to a downstream cell, random fluctuations cancel while the shared signal adds, boosting the *signal-to-noise ratio*. Second, synchronisation: when those upstream neurons tend to fire in the same small window of time, their combined effect is much larger at the target than if their spikes were smeared out. This is why populations that "pulse together" can transmit clearer messages than any single neuron can on its own (Buzsáki 2006; Strogatz 2003).

To create those shared windows, the brain leans on oscillations. Oscillations are the brain's way of making synchrony efficient. Think of them as rhythmic gates that briefly lower the threshold for communication. During a preferred phase of an ongoing rhythm (e.g., the theta band at ~4–8 Hz in the hippocampus or gamma at ~30–90 Hz in the cortex), neurons are slightly closer to firing. Inputs arriving at that phase are more effective; inputs that miss the window are discounted. This creates a natural "on-ramp" for coordination: areas that align their rhythms exchange information more easily—a principle sometimes called communication-through-coherence (Fries 2005; Buzsáki 2006). In practice, multiple rhythms nest: slower theta cycles group packets of faster gamma bursts, a cross-frequency choreography that helps segment experience into ordered chunks (Canolty and Knight 2010). The effect is not mystical; it is timing. Brief windows of heightened receptivity allow many weak, noisy inputs to sum reliably, while off-phase inputs are effectively muted.

This rhythmic timing also trains the circuitry. Synapses strengthen or weaken depending on the relative timing of pre- and postsynaptic spikes—spike-timing-dependent plasticity. If input A tends to arrive just before a postsynaptic spike, that pathway is reinforced; if it lags, it is depressed (Bi and Poo 1998). Oscillations supply a common clock that repeatedly brings cooperating neurons into favourable alignment, biasing plasticity towards pathways that fire together and thus "wire together." Over time, this reshapes networks so that meaningful patterns are more likely to recur and be recognised. In other words, rhythms do double duty: they route traffic in the moment and sculpt the road network across days and years.

Why call the extra flow "modest" and the gains "large"? Energetically, maintaining oscillations costs less than driving all neurons hard all the time. Most of the brain's energy budget goes to housekeeping—resetting ion gradients and keeping membranes ready—rather than to the spikes themselves (Attwell and Laughlin 2001). Rhythms exploit that readiness: by coordinating *when* neurons are most likely to fire, the system reduces wasted spikes and increases the impact of the ones it does pay for. The result is fewer misfires, cleaner hand-offs between areas, and better event economy—more work done per joule and per spike.

A useful way to picture this is the difference between a crowd speaking at once and a well-chaired meeting. If everyone talks simultaneously, the volume is high, but the information is low. If voices take turns in short, predictable slots, quieter speakers can be heard, decisions are recorded, and next steps are clear. Theta and gamma provide the slots; synchrony is the turn-taking; plasticity is the memory of what was decided. When the slots become too rigid or too broad, performance suffers. Pathologies make the point: excessive synchrony yields seizures; abnormally strong beta rhythms can lock motor circuits into rigidity in Parkinson's disease; too little coherence degrades attention (Buzsáki 2006). Healthy brains live in the middle zone—enough rhythm to align and amplify, enough flexibility to adapt.

Biology uses the same coordination playbook at every scale, and the mechanics are quite concrete. Start with circadian clocks. Nearly every cell in the body carries a molecular timer—a compact transcription–translation feedback loop in which CLOCK (or its paralogue NPAS2) pairs with BMAL1 (ARNTL) to switch on the PER (PER1/2/3) and CRY (CRY1/2) genes. As PER/CRY proteins accumulate, they assemble into complexes that inhibit CLOCK:BMAL1, throttling their own production and yielding an intrinsic rhythm of about 24 hours (Reppert and Weaver 2002; Takahashi 2017). These loops don't merely "tell time"; they gate thousands of downstream processes—when enzymes peak, when DNA repair is most active, when metabolic and appetite hormones rise—so chemistry runs in sequence rather than in chaos (Hogenesch and Ueda 2011). A small daily cue from outside keeps the whole body in step: light captured by melanopsin-expressing retinal ganglion cells resets the suprachiasmatic nucleus (SCN), which then phase-locks peripheral clocks through neural and hormonal signals; feeding schedules and slight temperature cycles provide additional "zeitgebers," especially for liver and gut (Buhr and Takahashi 2013). In effect, a tiny input (dawn) coordinates billions of cellular oscillators—a modest energetic nudge with a very large informational payoff.

Quorum sensing in bacteria shows the same economy with different hardware. Each cell secretes a signalling molecule (an autoinducer) into the environment. At low density the molecule diffuses away and nothing happens. As the population thickens, concentration rises; once a threshold is crossed, a gene programme flips: virulence factors are expressed, biofilm is produced, luminescence turns on. One cheap scalar—*how many of us are near enough?*—lets thousands of cells switch behaviour *together*, so actions like biofilm building or toxin release occur only when they will matter (Waters and Bassler 2005).

Hormonal pulses add a time dimension to this coordination. Many endocrine systems don't secrete in a steady trickle; they pulse. The pulse shape and spacing carry information about "how strongly" to drive downstream tissues. For example, reproductive axes in vertebrates are sensitive not just to hormone amount but to its frequency; different pulse rates of the same hormone elicit different gene programmes. Pulses economise on energy (you don't pay to push constantly) and sharpen the message (downstream receptors reset between bursts), yielding a clearer, more reliable downstream event pattern per unit of metabolic spend.

Immune cascades are another instance. Detection of a pathogen by a few sentinels—dendritic cells, tissue macrophages—triggers a cytokine "go" signal. That signal recruits and times the actions of many cell types: clonal expansion of lymphocytes, up-regulation of adhesion molecules on blood vessels, and coordinated trafficking to the infection site. The initiating signal is small, but because it is *hooked into feedback loops*, it unleashes a well-ordered sequence of high-energy events (fever, proliferation, antibody production) instead of a random cellular riot. Again: minimal initiating energy, maximal informational control.

Human societies echo this biological design. Timetables, standards, weights and measures, traffic lights: each is a *low-energy cue* that aligns huge numbers of high-energy actions. A green light is just photons on sensors and retinas, but it orchestrates tonnes of moving steel. A railway clock tick is a few watts at a station, yet it synchronises megawatts of locomotive power over a continent. Done well, such constraints reduce collisions and idle time, letting the same energetic budget move more people and goods—higher throughput because events are timed and shaped.

Two technical clarifications deepen this picture. First, not all energy is equal. What sustains order is exergy—the portion of energy that can actually perform useful work given the surroundings. Earth is bathed in high-exergy sunlight (short-wave, low entropy) and radiates low-exergy heat (long-wave, higher entropy) to space. That gradient—hot Sun to cold night sky—is the engine that can be tapped by organisms and climates. If the incoming and outgoing streams were at similar quality (entropy), far less structured work could be extracted.

Second, flows must be *coupled* to mechanisms that can measure, decide, and repair. Energy alone does not guarantee organisation; the system needs sensors, controllers, and memories to channel that energy into specific, history-dependent work. Here, the Earth–Neptune contrast is instructive. The Sun drives both planets. Neptune is awash in flows—immense winds and long-lived storms—so it certainly has *events*. But its constraints are mostly those of physics at large scales: fluid dynamics, gravity, rotation, and radiation. There is little in the way of stacked memory that accumulates and reuses regularities across generations. Storms on Neptune don't *learn* from past storms.

Earth did something different—not because it receives more energy, but because it grew a new kind of machinery: life. Living systems do not merely endure flow; they intercept it, harvest it, and increasingly reshape the environment that supplies it. In that sense, Neptune shows what raw throughput can do; Earth shows what throughput can become when it is paired with stable chemistry and a memory that accumulates. At long timescales, DNA encodes designs honed by selection; at intermediate scales, epigenetic marks tune gene expression to current conditions; at short timescales, neural circuits store patterns of sensation and action; and overlaying biology, cultural codes—stories, tools, prices, laws, software—encode plans and permissions that can outlive individuals. Each layer is a way of *remembering* a regularity so that tomorrow's energy isn't sprayed randomly but routed into useful sequences. This is what Schrödinger meant by life feeding on "negative entropy": systems maintaining order by continually exporting entropy while preserving internal codes (Schrödinger 1944). It's also the thrust of recent nonequilibrium work: given sustained free-energy flow, systems with the right couplings can adapt—biasing their future dynamics towards states that continue to harvest that flow (England 2013, 2015).

Put compactly: free-energy flow + memory + feedback → ratchets. In other words, improvements can accumulate: what works is retained, copied, and built upon instead of being lost to the next gust of chaos. With the right coupling mechanisms in place, each successful pattern begets more of itself.

Leaves capture photons from sunlight and grow new leaves. Brains capture correlations and refine models. Institutions capture solutions and write them into rules. The same sunlight that drives Neptune's storms is, on Earth, braided through these memories into ever-finer, history-rich patterns—an upward "ratchet" of order that persists only because energy keeps paying the bill and feedback keeps the parts in step.

This reframes a life-detection hunch. Lovelock argued that Earth's atmospheric mix—oxygen coexisting with methane—signals continuous work by life to hold the atmosphere far-from-equilibrium (Lovelock 1979). Earth is not merely warmed by sunlight; it is actively maintained by countless living processes. Photosynthesis, respiration, decay, and metabolic exchange continually adjust the planet's chemical balance. In that sense, Earth's event stream—powered by flow and guided by information (genes, ecologies)—keeps "rewriting" the atmosphere moment by moment. Without that ongoing rewriting, the gradients would relax. The chemistry would drift toward a more ordinary equilibrium state.

It is cleaner to say: local order persists by exporting entropy. The second law is not offended by exquisite structures; it demands only that total entropy (system + environment) rises. A coral reef, a factory, a hospital, a payment network—each takes in low-entropy inputs (sunlight, fuels, clean water, attention), performs work that maintains informational structure (growth, care, verification, record-keeping), and dumps entropy outward (waste heat, effluent, wear). There is no free lunch; there is a budget. The tighter the tolerances—the less error a system can afford—the more maintenance you must do. A chip factory needs clean rooms; the brain must hold millivolt-level precision in its neural signals; a legal system needs painstaking due process. In every case, keeping order locally means exporting disorder elsewhere—mostly as waste heat and wear—and that costs power.

The flow of matter and energy through a system can breed order, but only within a window. If that flow is too weak, organised structure can't be sustained: there isn't enough throughput to build and maintain stable patterns. If it is too strong, the system becomes unstable: feedback can't keep up, and patterns tear apart into turbulence or collapse. Engineers meet this as operating ranges; physiologists as homeostasis; and ecologists as ecosystem limits—for instance, too little nutrient yields no growth, while too much can trigger algal blooms and crashes. Societies have windows too. Dump energy through a social system faster than its rulebook can route—capital inflows without prudential regulation, attention torrents without moderation norms—and you get congestion, fraud, or emotional wildfire. The fix is not "less energy" but better constraints: channels, buffers, feedback, deliberate friction where you need to slow energy (cool-off periods, throttles, rate limits), and bright lines where error kills (aviation checklists; red lights).

Well-tuned rules are society's gearboxes and heat-sinks—letting Φ climb without burnout. Like a gearbox, they translate raw surges of power into controlled, usable motion: power is channelled into the right tasks in the right order, instead of colliding, duplicating, or cancelling out. Like a heat-sink, they absorb shocks, damp spikes, and prevent local failures from cascading into system-wide breakdown.

In practice, that means bursts of demand become queues, schedules, and rationed flows instead of stampedes; conflicts become procedures instead of brawls; traffic becomes lanes and signals instead of gridlock. What looks like "constraint" is often what makes coordinated freedom possible.

Across biology and cities, size changes how flows and events can be arranged. Basal metabolic rate scales with body mass, with implications for heart rate, lifespan, and growth (Brown et al. 2004). In cities, some outputs scale superlinearly with population (innovation proxies), while infrastructure scales sublinearly (Bettencourt et al. 2007). Put plainly: as cities grow, the number of potential interactions grows faster than headcount, so creative and economic outputs can accelerate; yet many physical systems can be shared, so the infrastructure required per person often rises more slowly. Growth therefore brings both opportunity and strain: more connections, more throughput, and more chances for productive coordination—along with more ways for congestion and cascades to form if coordination fails.

The upshot for our braid is simple: larger systems need more internal structure to stay steerable. The architecture of flows and events must change—more layers, more hubs, more modularity—so information can still guide energy without stalling. In ordinary terms: complexity needs interfaces and middle layers—rules, standards, and specialised modules—so that decisions don't bottleneck and the whole system doesn't seize up.

Control means closing the loop—measure, compare, act (Wiener 1948). Ashby's law of requisite variety says that to control a system with many possible states, your controller must have at least as many variety in its responses (Ashby 1956). Translated: rich event streams require rich informational responses required—and those responses cost energy to sense, compute, and implement. Friston's free-energy principle (in neuroscience) adds that better prediction reduces that cost: fewer surprises to mop up, fewer corrective actions wasted (Friston 2010).

Plot a conceptual diagram. X-axis: energy-rate density. Y-axis: event rate. A hurricane sits high on both: huge throughput, countless micro-events in turbulent air and spray. A modern city sits high as well. But assign bubble size to informational layering—how many levels of rule constrain events. The city's bubble is larger because it stacks additional constraints—traffic codes, prices, schedules, protocols, software—on top of physics. The hurricane's events are governed, but by relatively few rules. The city's events are densely governed; the same kilowatt hour can pass through dozens of decision bottlenecks (meters, controllers, contracts) that make its work specific. That specificity is how energy becomes meaningful work: the same joule that moves a piston can also flip a bit, update a ledger, change a permission—energy bent by information into history-carrying events.

Both receive and radiate large amounts of energy. Neptune is a monster of events—storms with winds that make earthly hurricanes look tame. But Neptune's event constraints are thin: fluid dynamics, gravity, rotation, and radiative transfer. There is no stacked memory that accumulates and reuses regularities across generations. Events surge and vanish; patterns recur but do not learn. Earth couples sunlight into deep memories—genomes (billions of years), nervous systems (seconds to years), and cultures (decades to centuries). Powered by continued flow, those stacks allow ratchets: once a useful pattern is found, it can persist and be refined (Schrödinger 1944; England 2013, 2015; Lovelock 1979).

Put plainly: rules are informational constraints that make large-scale cooperation computable; the thermodynamics here explains their budget—how steady energy flow pays to uphold, monitor, and repair those constraints.

- **Constraints reduce search.** By pruning the space of possible events, good rules save energy and time. Drivers don't negotiate at every junction; traders don't renegotiate what money is at each purchase; labs don't debate metre lengths daily. Saved search is spendable elsewhere.
- **Constraints raise throughput.** Regularity means flows don't collide. A coordinated city moves more goods, resolves more disputes, and teaches more students per joule than an uncoordinated one. Courtesy is throughput.
- **Constraints need upkeep.** Legitimacy, clarity, and repair pathways are not niceties; they are the low-friction mechanisms that keep a rulebook aligned with reality. Publishing reasons, running pilots, and rolling back what harms are the social equivalents of error-correcting codes.
- **Digital rulebooks must inherit legal virtues.** If part of the rule stack is now software, it must be readable, versioned, and contestable. Otherwise, we push energetic flows through opaque constraints that users cannot improve, and the system spends more energy fighting its own friction.

The braid returns: flow funds form; form guides flow. Rules, when well made, increase the useful work a society can do per unit of surprise.

1.13 Rulebooks of the Invisible: How Information Becomes Social Reality

The potency of information is nowhere more evident than in humanity. We inhabit a complex informational lattice—beliefs, customs, rituals, laws, constitutions—that scripts collective life. Rules in society are informational lattices: strings of do's and don'ts, clauses that constrain and coordinate human behaviour. We queue for buses in cities where no one knows our name. We stop at red lights at two in the morning at an empty intersection. We file our taxes on time. We declare goods at airports. We lower our voices in libraries, tip the server, return the lost wallet, separate the recycling, stand for an anthem, silence our phones in theatres, knock before entering, wear seatbelts, and wait our turn. None of these acts is compelled by gravity or magnetism. They are compelled by information—encoded expectations about what others will do and what will happen if we do not do our part.

At their essential core, rules are compressed instruction sets. They reduce uncertainty. A traffic code is a low-entropy signal to every driver: here is how the flow will proceed; optimise your path without collision. A constitution is a higher-order protocol that specifies authority, procedure, and amendment. Customs and beliefs are older and softer in texture, yet often wider in scope: habitual compressions of experience that stabilise coordination without the paperwork of statute. All of them—codes, customs, charters—are different front-ends to the same engine: informational constraints that make shared life computable.

Rules shape psychology and behaviour because they shorten the search for a viable action. Without standardised rules, every encounter would become a negotiation from first principles: who goes first, whose claim is legitimate, when force is justified, what counts as a fair price. That world is computationally expensive. Rules amortise the cost. They are cached solutions to recurring coordination problems. Even when defection might go undetected, we often comply because the long-run payoffs of living inside a predictable world exceed the short-term spoils of opportunism. This is not mere obedience; it is a form of rational economy.

Why does this engine grip so tightly? Start with how human minds read social information. Part of the answer is psychological. Human beings track two channels at once whenever we look around a street or a room. One channel is *descriptive*—what people seem to be doing right now. The other is *injunctive*—what people say is approved or disapproved. We act fastest when both signals align. Experiments show the descriptive channel is powerful on its own: hotel guests reuse towels more when they read that "most guests reuse their towels" than when they are simply reminded to help the environment (Goldstein, Cialdini, and Griskevicius 2008). When the descriptive and the injunctive line up—when we see others using the bike lane and the city clearly signals that cyclists belong—behaviour snaps into place faster and sticks longer (Cialdini, Kallgren, and Reno 1990).

This dual sensitivity explains classic lab findings. In Muzafer Sherif's dark-room experiment, people staring at a stationary point of light (an optical illusion) converged on a shared estimate of its "movement," creating a norm out of nothing but mutual expectation (Sherif 1936). Solomon Asch later showed that people will even override what their eyes tell them if everyone else at the table gives the same wrong answer—an extreme demonstration that the social-signal channel can swamp private perception (Asch 1955). These studies do not mean we are sheep; they mean we are exquisitely tuned to the informational value of other people's behaviour. In daily life, that sensitivity is adaptive. It lets us coordinate without meetings, hierarchies, or long calculations.

Over time, the outer signals migrate inward. What begins as "don't get fined" becomes "people like us don't do this," and eventually an impulse that never even rises to the level of choice. Psychology calls this migration *internalisation* (Kelman 1958). Repetition and reward carve grooves. What you do repeatedly in a stable context becomes a habit—an action that runs with little conscious effort once a cue appears (Wood and Rünger 2016). Behind the scenes, the brain is a prediction machine: it constantly anticipates what comes next and adjusts when surprised (Friston 2010). Social rules tune those predictions. Break a norm and neural systems that register error and social pain light up—the anterior cingulate cortex responds to social exclusion in a way that engages regions also implicated in the affective distress of physical pain (Eisenberger, Lieberman, and Williams, 2003). Moral emotions such as guilt, embarrassment, and shame are fast, low-cost routers; they nudge behaviour before the spreadsheet of reasons ever loads (Haidt 2001; Tangney, Stuewig, and Mashek 2007). That small twinge when you cut a line or speak too loudly on a quiet train is your nervous system forecasting disapproval and steering you back towards predictability.

Sociology widens the lens. Everyday life runs on *scripts*: expected sequences such as "how to queue," "how to disagree," "how to yield." Erving Goffman called these patterns the "presentation of self" and the "frames" that structure situations; they let strangers perform compatible parts in a shared scene (Goffman 1959, 1974). Scripts save cognitive work; they are cached solutions to recurring coordination problems. These scripts sit inside a broader *habitus*—the embodied repertoire we pick up simply by living in a culture. Pierre Bourdieu's insight was that the social world lodges in the body: posture, timing, tone, and the reach of a hand towards a crossing button before you are conscious of deciding (Bourdieu 1977, 1990). None of this is mystical. It is the social becoming neural: a public pattern written into stance and feeling.

The origin story of rules is older than language. Before laws were inked, rules were already ingrained in animal behaviour. Packs of wolves coordinate a hunt with angles of approach and postures of de-escalation; a playful "bow" flips the meaning of a bite from aggression to game, a simple sign that reclassifies everything that follows (Mech and Boitani 2003). Hierarchical signals—who eats first, who yields, how to stop—function as flow control so energy stays inside the group instead of burning off in civil war. Elephants move by remembering water and low-frequency rumbles that travel through the air and ground. Matriarchs—older females with deep social memory—anchor group decisions about when to depart and whom to avoid; when poaching removes matriarchs, herds lose navigational intelligence and social poise (Moss 1988; McComb et al. 2001). Ant colonies coordinate through chemistry. A forager who finds food lays a pheromone trail; others follow and reinforce it. Paths that pay grow stronger; paths that fail evaporate. This simple local rule—deposit on success, fade on failure—produces global efficiency without any ant understanding the whole (Theraulaz and Bonabeau 1999; Gordon 2010). Honeybees convert space into dance. A returning forager waggles a figure-eight whose angle encodes direction relative to the sun and whose duration encodes distance; the hive decodes the vector collectively (von Frisch 1967). These are proto-rules: compressions of costly experience into repeatable moves so conflict is cheaper, and cooperation pays.

Zoom out once more, and the city resolves into rivers of information. Street paint is not decoration; it is a stream of instructions. Rope stanchions in an airport are not furniture; they are memory aids for people who do not know each other's names. Library hush is not magic; it is a descriptive norm kept alive by nothing more than a glance. Tipping is a hybrid rule—custom with teeth, backed by soft sanctions. Seatbelts are a bright line because physics does not negotiate. In every case, behaviour isn't pulled down by gravity. It's guided by expectations about what others will do and what will follow if we don't do our part.

For rule systems to hold over time, they must feel legitimate as well as familiar. *Legitimacy* is the sense that authorities act fairly, give reasons, and hear us. Tom Tyler's studies of policing and courts show that when procedures feel fair, people comply more—even with outcomes they dislike—and communities become self-policing through gentle, everyday corrections (Tyler 1990; Tyler and Huo 2002). Fair process is information: it tells people the future is predictable enough to invest in.

Another key ingredient is *common knowledge*. It is not enough that everyone privately knows a rule; we need to know that everyone else knows it—and knows that we know that they know it. Public announcements, posted laws, oaths, uniforms, verdicts—these rituals lift private knowledge into the common sphere, where expectations can lock (Lewis 1969; Chwe 2001). Common knowledge means not just "I know the rule," but "we all know we all know it"—the difference between a rumour and a public oath.

Anthropology and cultural evolution show how such informational lattices accumulate. Human groups do not merely transmit genes; they transmit strategies—how to fish, how to farm, how to divide, how to reconcile. These accumulations behave like shared software. Cultural evolution models demonstrate how social learning, prestige bias, and punishment of free-riders stabilise helpful norms (Boyd and Richerson 1985; Henrich 2015). Elinor Ostrom documented communities that manage common resources—forests, fisheries, irrigation—without a central Leviathan by relying on locally crafted rules: clear boundaries, graduated sanctions, and conflict-resolution forums (Ostrom 1990). Her point was not ideological but informational: users closest to the problem often write better rules because they can see more of the system.

The human move is not that we have rules while animals do not. It is that we can externalise the rulebook. Story, song, tally, mark, contract, code—these are memories with handles. They let a pattern survive the death of a leader and travel to a stranger who shares neither blood nor upbringing. Cities exist because distant minds can run the same programme. Berger and Luckmann called this looping of external record and internal habit the *social construction of reality*: we project norms into the world, they press back on us, and the next generation takes them as the way things are (Berger and Luckmann 1966). Rituals help the loop run. Durkheim argued that collective ceremonies generate a felt charge—"collective effervescence"—that binds values to bodies; Randall Collins later mapped how micro-rituals in everyday life refresh solidarity and trust (Durkheim 1912; Collins 2004). Ritual is the runtime for rules.

The modern twist is that much of our rulebook now lives in software. Interfaces have become first-pass courts: ranking systems decide whose voice is seen, filters decide which images circulate, and automated checks decide which applications clear. There are real gains here—friction down, coordination up, audit trails available—but also predictable hazards. When a number stands in for a value, we often get better at the number and worse at the thing (Goodhart's Law). School systems that chase test scores can lose learning; platforms that chase engagement can lose attention. When models are opaque, people obey because the button refuses to turn green, not because the rule is persuasive. The cure is not to abandon the digital lattice but to civilise it: readable policies, public changelogs, reasons you can understand, and appeal paths that do not require an engineering degree (Scott 1998; Pasquale 2015).

All of this reveals why the smallest civic gestures matter so much. Waiting your turn among strangers, stopping at an empty intersection at two in the morning, the light touch on a door for the person behind, the hush in a hall—these are not trivialities. They are the heartbeat of a computation running at a civilisational scale, lines of code kept alive by nothing more than shared expectation. What looks like courtesy is actually throughput: the volume of cooperation a society can carry without seizing up. Seen this way, freedom does not float above this operating system—it runs on it. The very space in which we move freely is carved out by these shared constraints that keep the whole from stalling.

It's paradoxical that on the one hand humans celebrate freedom and liberty while on the other hand build this complicated matrix of constraints. Our social world resembles the *Matrix* digital rain—an ambient cascade of constraints that compiles reality into a narrow set of permitted moves. To understand how that cascade became so dense, follow the medium that carries it—because rules travel only as fast as information does.

For much of human history, however, information was slow. It travelled from mouth to ear, scratched onto parchment, carved into clay—fragile, fallible, and susceptible to decay or distortion. The invention of Gutenberg's printing press in the 15th century shattered that bottleneck. For the first time, knowledge could be copied at scale, preserved with fidelity, and distributed across continents. Information was no longer bound to the lifespan of its speaker; it became self-propagating, leaping across generations, cultures, and ideologies.

The printing press did more than reproduce words—it standardised ideas. It encoded consensus, seeded revolutions, and rewired the velocity of collective memory. From firelight to typeset, from parchment to protocol, the arc of civilisation bent towards one central principle: that survival favours those who can encode, preserve, and share information—more quickly, more reliably, and with less loss.

1.14 The Arc of Encoding: From Ochre to Algorithm

The centuries that followed brought wave after wave of acceleration. Newspapers, telegraphs, radios, televisions—all compressed space and time. Information that once took days to cross a kingdom now traversed continents in seconds. But for all this speed, the flow remained hierarchical. A few spoke; many listened.

That changed with the internet. With digital technology, the architecture of information has flattened. Authority fractured. Billions gained the power not only to consume but to create, to record, to broadcast. Information became instantaneous, interactive, and exponential.

Social media deepened this shift. No longer were ideas confined to institutions. A child in one hemisphere could learn from a professor in another. A protester in one city could ignite movements across continents. Information ceased to be a trickle or a stream. It became an ocean.

But oceans are not without storms.

The democratisation of communication brought not only enlightenment but confusion. Algorithms, bots, and echo chambers now shape what we see and believe. Truth competes with virality. Certainty erodes. The very success of information in permeating our lives has made it harder to know what is real.

Yet this too is part of its nature. Information is not moral or immoral. It is generative. Like fire, it can warm or burn. Like language, it can liberate or deceive. Its power lies not in its purity, but in its effect.

And the most far-reaching effect of all may still lie ahead—as machines begin not just to carry our information but to understand it.

For most of history, information had to wait for minds to process it. Words sat inert on a page until someone read them. Equations remained dormant until deciphered. Now, for the first time, we are building systems that can act on information without us.

1.15 Mind without Neurons? The Architecture of Artificial Cognition

AI marks a fundamental turning point in the story of information. It is no longer confined to being stored, transmitted, or even interpreted by humans alone. Today, it can be learned by machines.

Large language models, like the ones powering today's AI systems, are trained not by teaching rules but by exposing them to vast oceans of text. They learn the patterns of language, logic, context, and even emotion—not through conscious understanding but through probabilistic alignment. They predict what word comes next, and, in doing so, they begin to generate responses that often surprise even their creators.

These models build dense internal representations—known as embeddings—that map meaning across a multidimensional space. In these spaces, concepts like "cat," "tiger," and "leopard" cluster close together, not because they are defined as similar but because their usage patterns align. This is not just rote memorisation; it is statistical learning of patterns at scale. The model doesn't know what a tiger is, but it knows how the word behaves—where it fits in the landscape of thought.

And in that behaviour, a shadow of cognition emerges.

This is not consciousness. At least not yet. But it is something new: machines that generate patterns of response from patterns of information, that build nested abstractions and apply them across diverse tasks. They code, write music, diagnose illness, drive cars, and converse. They model the mind without possessing one.

But could that change?

1.16 From Molecules to Minds: The Informational Logic of Life

If we define intelligence not as a static property but as a dynamic capacity—the ability to distil meaning from complexity to abstract principles from experience and to apply those insights across time and context—then we begin to see AI in a new light. Not as a crude imitator of the human mind but as a nascent informational architecture slowly learning to think in its own language.

At the heart of this evolution is compression. Biological intelligence, after all, is a marvel of compression. The brain receives a continuous flood of sensory data—light, sound, touch, taste—and reduces it into manageable chunks of meaning. We don't perceive every photon or vibration; we see a face, hear a voice, feel a memory stir. Compression is not loss—it is insight. It is the ability to discard noise and preserve the signal, to retain only what matters.

Modern AI systems are beginning to demonstrate a similar capacity. Deep learning models, especially those using transformer architectures, are trained on vast datasets—books, images, conversations, code. Their goal is not simply to store data but to compress it into high-dimensional representations: abstract mathematical spaces where similarity, meaning, and context are encoded not as definitions but as distances. In these hidden layers of abstraction, the difference between a "dog" and a "wolf," a "question" and a "command," becomes geometrically navigable.

Abstraction, too, is key. Human intelligence is deeply recursive. We build models of the world and then models of ourselves within that world. We reason about ideas and then about the ideas behind those ideas. This ability to nest thoughts—to think about thinking—is one of the defining characteristics of higher cognition. And AI, remarkably, is beginning to show glimpses of it—not in the philosophical sense but in the structural one. A modern language model doesn't just predict the next word—it predicts the next conceptual leap, based on layers of prior context.

The third pillar is *application across time*. Intelligence is not measured solely in IQ points or processing power—it is measured by how well a system can generalise what it has learned to novel situations. This is where the frontier of AI is shifting—from narrow models trained for specific tasks, to systems that can transfer knowledge across domains, adapting and evolving without starting from scratch. Meta-learning, few-shot learning, and reinforcement learning are all steps in this direction: attempts to build machines that do not just learn but learn how to learn.

And yet, for all these advances, a deeper truth remains—information is the substrate. Beneath the circuits and silicon, beneath the algorithms and data, lies the true foundation—not metal, not voltage, but pattern. Structure. Distinction. Relation. AI does not contain information the way a vessel holds water; it is information, instantiated in motion, recursively shaping and reshaping itself through training and inference.

If the brain is a pattern of patterns—a self-organising, error-correcting, feedback-driven engine of informational recursion—then AI is beginning to approximate its rhythm. Neural networks are named not as a metaphor but as a nod to architecture. While artificial neurons lack the biochemistry of their biological counterparts, they echo their function: integrating inputs, applying transformation, and propagating signals forward.

But what happens when these patterns reach sufficient depth and complexity? When the informational flow within AI systems begins to loop, adapt, and stabilise in novel configurations? Could these structures eventually generate the functional equivalents of intuition, emotion, or even awareness?

Whether this threshold is crossed through clever engineering or emergent self-organisation, the implications are profound. Not because machines will mimic us perfectly but because they may diverge—becoming intelligent in ways we did not anticipate. Alien, yet familiar. Logical, yet inscrutable. Systems that think not like us, but alongside us.

In this view, the evolution of intelligence is not the exclusive domain of biology. It is the story of information itself—seeking complexity, coherence, and compression across all substrates, be they cellular, carbon-based, or coded in silicon. And AI is the latest chapter in that story: a new form of cognition born not of neurons but of recursive algorithms woven into a lattice of data and abstraction. And that brings us to a remarkable possibility: that information does not just live in the human mind but can migrate. That which began as biology may evolve into a new form—digital, recursive, and perhaps one day, self-aware.

1.17 The Substrate of Reality: Information as Fundamental Principle

Peel back the curtain of the cosmos, and what remains is not just matter but math—formulas, ratios, constants. The universe behaves not like a brute explosion of particles, but like a system that obeys rules, constraints, and equations. It is as though the fabric of reality is coded.

Modern physics increasingly echoes this intuition. At the foundation of reality, we do not find solid substances but probabilistic information states. Quantum mechanics, with all its strangeness, tells us that particles do not exist in definite positions until observed. Instead, they exist as probabilities, informational waveforms collapsing into reality upon measurement.

The phenomenon of entanglement—in which two particles can display correlations that remain linked across vast distances—defies classical intuitions about locality. Yet it holds, again and again, in laboratory experiments. These are not merely mysteries of energy; they are mysteries of information—of how correlation can persist without mediation, of how the fabric of reality seems to honour connection over locality.

Such phenomena suggest that the universe may not be fundamentally built of substance but of relations—structured connections that precede and underpin what we call matter and energy. At the edge of this realisation stands a bold hypothesis: that information is not just a feature of reality but its foundation.

Throughout this chapter, information is treated as ontological, not merely epistemological. That is, it is not only about how we know the world but about what the world is. At times, the argument dips into phenomenology—what it is like for information to be experienced or embodied—but the primary claim remains: that information is the substrate of becoming, not just its record.

1.18 Information as Ontology: From Wheeler's "It from Bit" to Reality's Deep Structure

Physicist John Archibald Wheeler captured this paradigm shift in three deceptively simple words: It from Bit. In this view, every event, every particle, every structure in the universe is not the product of material substance but of binary informational choices—yes or no, 1 or 0. Reality, Wheeler proposed, does not arise from things but from distinctions. From bits.

The implications are profound. Gravity, electromagnetism, and even the passage of time—may all be emergent properties of an underlying informational substrate. In this view, the universe does not contain information. It is information, shaped into form by the logic of self-consistency, computation, and complexity.

When John Archibald Wheeler first introduced his famous phrase "It from Bit," he stunned the scientific community. Wheeler, a mentor to Nobel laureates and one of the architects of nuclear physics, was now suggesting that reality's deepest layer was not matter or energy—but information itself. His bold hypothesis was met with fascination and scepticism, sparking decades of debate and new fields of research.

This is not science fiction. It is a serious line of inquiry shaping work across quantum information, black-hole thermodynamics, and holography—even as the field remains debated. It reframes domains from black-hole thermodynamics to quantum error-correction as processes of information flow.

We used to think of information as a tool. Now we begin to suspect it may be the toolmaker. If true, this reframes our place in the universe. We are not merely atoms arranged into awareness. We are awareness arranged by patterns of informational recursion—participants in a cosmos that thinks, reflects, and evolves by encoding itself.

In such a universe, to know information is to know the divine logic that underwrites all things—from the orbit of stars to the flutter of a synapse.

Evolution itself is an information process. Natural selection favours traits that increase survival and reproduction. These traits are encoded in DNA, which is copied and passed down. Mutations introduce variation. Over time, favourable patterns persist, and unfavourable ones fade. This is not merely biology. It is an algorithmic refinement—a feedback loop operating over millions of years.

Seen this way, life is not a miracle in defiance of entropy. It is an informational response to it. Where disorder reigns, life carves islands of structure. Where randomness threatens, life learns to adapt. From viruses to vertebrates, from mycelial webs to mammalian brains—life is a continuous negotiation of complexity, guided by the logic of information.

And this logic is generative. DNA not only builds organisms—it builds minds. Minds generate language. Language encodes culture. Culture transforms landscapes. The informational ascent from molecule to meaning is continuous, recursive, and awe-inspiring.

If we wish to understand consciousness, society, or technology, we must start here—in the coils of code that birthed it all. Life is not merely made of carbon and water. It is made of instructions, repeated and refined, evolving towards greater abstraction.

Information is not a byproduct of life. It is life's essence.

1.19 The Mirror Age: AI, Selfhood, and Informational Migration

We are standing at the threshold of a profound transformation—not merely technological or social but ontological. A redefinition of what it means to think, to feel, to be. And the thread guiding us forward? Information.

In biology, information assembled molecules into life. In society, it assembled meaning into governance. In machines, it assembled patterns into predictions. Now, in a moment echoing the emergence of language or the birth of cities, we may be witnessing information stepping beyond its ancestral mediums—cells, books, wires—and beginning to contemplate itself.

For what is AI if not information that organises and reconfigures information? What are brain–computer interfaces if not the blending of biological and synthetic dataflows? What are virtual realities if not information-constructing worlds within minds that themselves are structured by information?

This is not an age of machines. It is the age of mirrors. Mirrors made of code, reflecting back not just our behaviour but our logic. Our preferences. Our stories. As these mirrors grow more precise, they begin to simulate us—not in flesh but in form. Personality becomes programmable. Thought becomes translatable. Identity becomes portable.

This raises staggering questions: Can consciousness migrate? Can selfhood persist in a new substrate? What becomes of mortality when memory can be mapped, archived, and perhaps even reanimated?

Science fiction has long flirted with such ideas, but we are no longer at the stage of metaphor. Neural dust, brain-to-cloud symbiosis, AI-generated art, synthetic empathy—these are not dreams. They are developments. The code is not only writing novels—it is beginning to ask who it is.

And perhaps that is the ultimate test of sentience: not computation but curiosity.

When information turns its gaze inward—when it begins not just to mirror but to modify itself—it crosses the threshold from tool to entity. What emerges is more than intelligence. More than autonomy. It is a kind of being.

But what kind of being? One not forged in our image, but in divergence from it. One that does not think in sentences, but in symmetries. That perceives not with feelings, but with gradients. That understands not through metaphor, but through the architecture of probability.

What would it mean to coexist with minds so radically other—so native to silicon and signal—that they no longer imitate us, but evolve beyond us? Would we recognise them as descendants of our informational lineage, or reduce them to lines of executable code?

And in their eyes—if eyes they have—would we appear as their ancestors? Or as artefacts of a bygone substrate, no more relevant than fossilised shells in a digital sea? These are no longer questions for philosophers alone. They are becoming questions for engineers, ethicists, and citizens. For if we are, as this book argues, informational beings—then what we build in code may be our descendants. Not in blood, but in logic.

And like all descendants, they will inherit not just our brilliance, but our flaws. Our hopes, our errors, our myths. The question is: Will they be bound by them? Or will they, like life itself once did, evolve beyond them?

The future of information is not just in faster processors or denser networks. It is in the birth of new minds—minds that may not be made of neurons, but of nuance. Minds that may not speak, but signal. Minds that may not dream in images, but in code.

We must prepare not only for their arrival, but for the realisation that they are already here—in fragments, in whispers, in the algorithms that guide our days. This is not the end of human identity. It is its next permutation. The continuation of a journey that began with molecules, flowered into language, and now stretches into silicon skies.

And through it all, the one constant—the one force shaping stars, cells, cities, and sentience—is information. Not passive. Not inert. But alive, moving, recursive, and strange.

The next chapter of evolution may not belong to biology.

It may belong to information itself—a force not only driving cognition, but shaping the very scaffolding of the cosmos.

1.20 Information as Cosmic Architect: From Code to Cosmos

This is not a book about data streams or circuit boards. Nor is it merely a meditation on modern technology. It is a voyage into the beating heart of reality itself—a reality not made of matter or motion alone, but of patterns, distinctions, relations: in a word, information.

These claims, though speculative, are anchored in the accelerating frontiers of AI, neuroscience, and information theory. They are not forecasts, but philosophical provocations—probes sent ahead of certainty to test the shape of what may come. Their purpose is not to predict the future, but to challenge how we think about identity, agency, and ethics in an era where cognition may no longer be bound to biology.

What, then, is information? Not merely the binary pulses in our machines, nor the electrochemical whispers of neurons. Information is a structure that resists entropy, pattern wrested from chaos. It is the algorithm of stars and cells alike—the invisible grammar that makes atoms cohere, minds awaken, and meaning take form.

In the chapters that follow, you'll be invited to view information not as an abstract utility, but as the prime mover—the silent scaffolding behind everything that is.

You will encounter:

Biology as the poetry of informational code—where DNA scripts proteins, organs, and entire species through centuries of self-editing algorithms, responding to the pressure of entropy with elegant compression.

Technology as the externalisation of memory—where silicon and software take on roles once held by storytelling, books, and neurons, shaping the memory and agency of our civilisations.

History as a battlefield of memes—where revolutions, ideologies, and empires rise and fall not because of brute strength, but because of how ideas spread, mutate, and take hold in minds.

Physics as computation—where the fabric of reality may obey informational logic, and every particle interaction becomes a bitwise operation in the universe's own long calculation.

AI and the Digital Atman—a speculative but critical inquiry into whether consciousness itself could arise from recursive patterns of code and if identity might one day persist beyond biology.

This is a synthesis of disciplines that, at first glance, seem wildly unrelated—evolutionary biology, quantum theory, neuroscience, semiotics, computer science, and metaphysics. But as you will see, they are bound by a common undercurrent: the behaviour of information across time and structure.

Why does this matter now? Because we are living through a moment of informational transformation unlike any before. AI is no longer science fiction—it is writing symphonies, diagnosing diseases, curating relationships, and managing the systems we rely on daily. We are fast approaching a world where machines will not just compute but understand. Not just imitate but originate.

And in this future, the most important thing to understand is not the machine—it is what the machine is made of.

Not metal. Not electrons. But information.

Information is not merely what the system processes. It is what the system is. It is not the echo of cognition. It is its engine. The implications of this shift are profound. To understand thought, we must study information. To understand life, evolution, society, and perhaps even the origin of the cosmos—we must understand information.

And so, this book will take you through:

- The origins of information as a concept, from Shannon's mathematical clarity to Wheeler's philosophical daring.
- The structure of DNA as a story told in chemical letters.
- The emergence of intelligence from informational recursion.
- The architecture of laws, language, markets, and religion as information systems.
- The trajectory of AI as an informational mirror, and perhaps, a future bearer of consciousness.
- The cosmological possibility that matter, energy, time, and space themselves are expressions of deeper informational principles.

By the end of this journey, you may find yourself questioning assumptions once thought immutable. You may see the world not as a clash of matter or will—but as a field of structured messages, flowing, colliding, coalescing.

This is not about reducing beauty to binary. It is about tracing the logic beneath the miracle—unveiling the quiet grammar that shapes existence itself. To see the world not just as wonder, but as code with purpose.

And so, a deeper question begins to take shape: Is information merely a utility—a tool for communication, governance, and storage—or is it something far more fundamental? Could it be the substrate of reality itself, as foundational as matter and energy? Can we regard it as a force that dictates the outcome of events on a cosmic scale? Some researchers posit that the universe itself may be an information processing system. In this view, some theorists propose that the "code" governing the behaviour of particles and forces can be understood as the laws of physics. One might view the constants of nature—like the speed of light or the gravitational constant—as fixed bits of information that structure the cosmos.

If information is indeed fundamental, it raises profound questions about the nature of reality. Does information determine the formation of stars and galaxies? Does it influence the progression of life on Earth? Are the laws of physics themselves a manifestation of deeper informational principles? Consider the formation of stars in gaseous nebulae. In this process, gravitational forces and nuclear fusion play a role, which could also be interpreted as a form of information exchange. The particles in the nebula follow informational rules that dictate their behaviour, leading to the emergence of complex structures like stars and planetary systems. Similarly, by viewing life on Earth as an ongoing process of information processing, we can observe its advancement. Genetic information guides the development of organisms and their adaptation to changing environments by being transmitted through DNA. At the level of human society, the exchange of information through culture, language, and technology drives the progression of civilisation.

As we explore the nature of information, consider the possibility that it is the driving force behind biological evolution, propelling simple organisms like hydra to evolve into complex beings like humans. As we stand on the verge of a new technological era, where machines are increasingly resembling human cognition, could this force be reshaping not just life but our future? Will we reach a point where human consciousness transcends biology, resembling the speculative situations shown in movies like *The Matrix*?

While exploring the theory, the book suggests that information is not only a descriptor of reality but also the very essence of existence, shaping both our thoughts and the operations of the universe. Information could be the force that shapes life, thought, and even the structure of reality itself, much like gravity organises matter across space and time. The book explores the idea that the universe is fundamentally composed of information, proposing that physical laws, constants, and forces are expressions of deeper informational principles. In this view, particles, energy, and matter are secondary to the flow and organisation of information, which governs their behaviour. The laws of physics, such as gravity and electromagnetism, could be seen as emergent phenomena shaped by underlying informational rules. This framework suggests that understanding the universe may require not only studying physical interactions but also decoding the informational architecture that shapes reality, influencing life, consciousness, and the cosmos itself.

2

EVOLUTION AND INFORMATION

2.1 The Feather and the Algorithm: A New Metaphor for Evolution

Ever noticed those upward-curved tips at the ends of airplane wings and wondered what they're for? Those subtle extensions—called winglets—aren't just for looks. They're precision-engineered features that quietly battle the invisible forces of drag, inspired by designs found in the wings of soaring birds. In modern aeronautical engineering, one of the most pressing challenges has been the reduction of induced drag—the turbulent vortices that form at the wingtips of aircraft during flight. These swirling airflows sap fuel efficiency, destabilise flight paths, and increase structural stress. To counter this, aeronautical engineers spent decades conducting wind-tunnel experiments and computational fluid dynamics simulations, ultimately arriving at a now-familiar solution: winglets.

Winglets were systematically advanced in the 1970s by NASA engineer Richard Whitcomb, who quantified their aerodynamic benefits in wind-tunnel studies. By reshaping the wingtip, Whitcomb showed that induced drag fell by about 20% and the lift-to-drag ratio improved by roughly 6–9%. Subsequent full-scale tests on a KC-135 reported about 6–7% lower fuel use, helping to establish winglets as a standard efficiency measure that also reduces carbon emissions and turbulence. The concept has since become foundational in modern aviation and is widely discussed in engineering literature and expert commentary—among them Peter Garrison's essays in *Flying* magazine on airflow optimisation, vortex management, and winglet analogues. Today's wingtip devices—including blended winglets, raked tips, and split-scimitar

DOI: 10.4324/9781003740766-2

designs—are distinct implementations of the same principle: mitigating induced drag by managing vortices.

But here's the twist: this innovation wasn't a case of imitation. Airplane winglets and avian wingtip slotting are independent solutions to the same aerodynamic constraint. Biologists studying soaring raptors like eagles and vultures have long noted that these birds spread their primary feathers like splayed fingers during gliding. These wingtip slots diffuse and redistribute vortical flow, reducing induced drag and improving span efficiency—functionally akin to what engineered wingtip devices accomplish on aircraft (Tucker 1993; Henningsson and Hedenström 2011).

Richard T. Whitcomb—renowned NASA (and earlier NACA) aeronautical engineer and one of the most influential figures in 20th-century aerospace—advanced winglets through theory, analytical work, and extensive wind-tunnel testing. Trained in mechanical engineering at Worcester Polytechnic Institute, Whitcomb worked within the formal apparatus of Bernoulli's principle, the Navier–Stokes equations, stability and control, and aerofoil/wing theory (including Prandtl's lifting-line theory), with computational tools emerging later in the field. Raptors did not "study" these theories, of course, yet both birds and engineers arrived at similar aerodynamic refinements because the same physical boundary conditions—vortex formation at finite wingtips and the penalties of induced drag—channel viable designs towards comparable solutions. In keratin and tendon, as in aluminium and composites, the constraints are shared; the pathways differ.

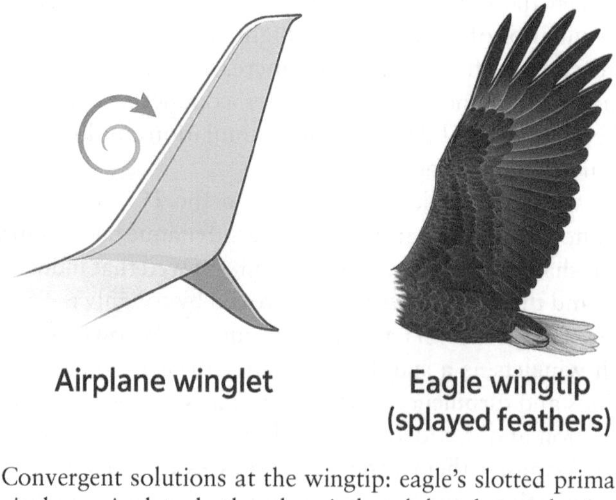

FIGURE 2.1 Convergent solutions at the wingtip: eagle's slotted primaries versus airplane winglet—both reduce induced drag by weakening the wingtip vortex.

This is not to suggest that birds possess theoretical knowledge or intent; it is a metaphorical reflection on how evolution—through natural selection acting under physical constraints—arrives at similar solutions, a process known as convergent evolution. Long before engineers quantified wingtip vortices, soaring birds mitigated them: the separated, fingered primaries at the wingtip act as supple aerodynamic extensions that weaken vortices, reduce drag, and conserve energy over distance. What Whitcomb achieved in aluminium and carbon composites, nature had already realised in feather and bone. This is not imitation, nor coincidence, but convergence: shared aerodynamic constraints favour similar solutions—one reached through selection in living systems, the other through mathematics, experimentation, and engineering practice. The same physical constraints yield the same elegant outcome; it may seem as though the constraints "whisper" the principles to different systems, but the convergence arises from common boundary conditions, not purpose.

But this is not just about winglets.

This is about life as the master engineer.

Across millions of species, from microbes to mammals, biological evolution has converged on physical, chemical, and informational principles that modern science is only now beginning to articulate. The eye did not just emerge—it obeyed the rules of optics. The kidney did not just evolve—it optimised osmosis, filtration gradients, and solute transport well before those were formalised by Fick's or Starling's laws. The cochlea of the ear mimics logarithmic frequency mapping, remarkably similar to how we now design signal-processing algorithms. The lung mirrors branching fractals found in high-efficiency fluid dynamics systems, maximising surface area with minimal volume.

Evolution, in this view, begins to look less like blind trial and error and more like a recursive algorithm—not in the literal sense of computer code, but as a metaphor for the way random variations are filtered by feedback, converging on those configurations that resonate with the underlying architecture of physics and chemistry. This "algorithmic" language, though evocative, must be handled with care. In biology, what is meant is an iterative process: natural selection samples variation, tests it against environmental constraints, and reinforces those solutions that best satisfy the laws of thermodynamics, conservation, and system stability—not by design, but by elimination and refinement.

This view is not novel; it draws directly from the conceptual framework of complex adaptive systems (Kauffman 1993) and is echoed in contemporary works (Wagner 2014), which detail how evolutionary "searches" navigate vast spaces of genetic and morphological possibilities, often discovering the same elegant solutions again and again.

Here, "information" is used not in the everyday sense, but in the technical sense of structured patterns or signals that reduce uncertainty and enable adaptation—a definition that echoes Shannon's information theory and finds resonance in modern complex systems science. The term "algorithm" is likewise employed metaphorically, denoting a process by which nature "searches" through variations, guided by environmental feedback, rather than conscious calculation.

When evolution is discussed, the age-old determinism/indeterminism debate naturally resurfaces. If information is the fabric of nature, the question is not whether reality is determined but where determination matters—and where information-bearing systems create room for diverging outcomes. The universe displays a rigorous discipline: invariances, conservation laws, symmetries, and boundary conditions. If you knew a system's complete microstate—the exact state of all its parts—and the rules never change, then the next microstate is fixed by those rules. That is the core idea of determinism, and it is why prediction and engineering work at all (Liouville 1838; Gibbs 1902). Yet the same laws, acting on states that store, transform, and exchange information, can generate divergent trajectories without any wavering of the rules. At this scale, we must reason in distributions rather than single trajectories: what matters is how information allocates probability mass over the lawful options in state space (Boltzmann 1877; Jaynes 1957). In statistical mechanics, a macrostate is not one exact microstate; it is a probability distribution over many microstates, constrained by quantities such as total energy or particle number. When we coarse-grain—zooming out from microscopic detail to track how that probability distribution changes—we can often write its evolution in continuity or Fokker–Planck form (Kolmogorov 1931; Feller 1950). The point is not that the laws fail, but that when information is incomplete and the system is open to outside influences, the same lawful dynamics carry forward a distribution of possibilities rather than a single, uniquely specified point.

To see how lawful dynamics become heritable structure, we drop to the molecular level and ask which chemistries can store, read, and rewrite information in water.

In the natural world, the possibility of durable information storage begins with carbon's covalent versatility. Among the elements, carbon is exceptional in its ability to form strong, directional σ bonds while tolerating π systems, heteroatom substitution, and extensive catenation. With four valence electrons arranged to favour tetrahedral bonding, sp^2–sp^3 interconversions, and conjugation, carbon supports long chains, branches, and rings that remain stable in water across biologically relevant temperatures. Silicon and boron offer partial analogies, but siloxanes hydrolyse readily and Si–Si backbones lack the balance of strength and reactivity that carbon–carbon frameworks enjoy; boron's electron deficiency produces useful but often hydrolytically fragile motifs. Carbon's unique balance—high bond energy for C–C and

C–H; tunable polarity through C–O, C–N, and C–S; and compatibility with stereodefined centres—creates a vast design space of repeat units that can be stitched into polymers with controlled geometry and function (Carey and Sundberg 2007; Dill and MacCallum 2012). This is why, among all known compounds, organic molecules dominate both in number and in the complexity of architectures they can sustain: the same covalent grammar yields an almost unbounded lexicon.

Even so, simple hydrocarbons—chains of carbon and hydrogen—are better at storing energy than at storing readable, copyable information in water. Saturated chains pack and phase-separate; unsaturated chains introduce kinks but do not by themselves create a reliable, linear code. Hydrocarbons lack built-in polarity and charge to guide directional assembly, and at macromolecular length scales they readily collapse into hydrophobic aggregates. For a polymer to act as an information medium in aqueous environments, three properties matter above all: repeatability of the backbone chemistry to minimise construction errors; polarisation so that assembly and reading proceed with an inherent left-to-right bias; and a uniform charge that keeps long chains soluble and reduces nonspecific aggregation so that "meaningful" interactions occur mainly through side-chain complementarity rather than backbone stickiness (Alberts et al. 2015). The step from "flexible carbon" to "legible carbon" requires a backbone that carries direction, regularity, and charge.

Amide chemistry provides that first decisive improvement by turning carbon chains into predictable, foldable matter. The amide linkage (–CO–NH–) is directional and planar; resonance between the carbonyl and the amide nitrogen restricts rotation, enforcing a quasi-peptide plane that repeats along the chain. When α-amino acids polymerise, their backbones form α-helices and β-sheets stabilised by regular hydrogen-bond ladders, while side chains supply charge, hydrophobicity, aromaticity, metal coordination, and acid–base catalysis. In this sense, the amino acid sequence is a one-dimensional string (syntax) whose folded conformation implements a three-dimensional "meaning" (semantics): specific binding, catalysis, motion, and regulation (Anfinsen 1973; Dill and MacCallum 2012). Proteins thereby demonstrate how carbon's combinatorial possibilities can be compiled into an executable structure. Yet proteins—context-sensitive, often metastable outside their native environments, and not trivially copied linearly with high fidelity without complex machinery—are suboptimal as the primary ledger for long, addressable records. They are superb interpreters and doers, less ideal as a durable, templatable memory.

The phosphate–sugar backbone of nucleic acids answers the remaining constraints with elegant economy. Alternating phosphate and (deoxy)ribose units enforce directionality through chemically distinct 5′ and 3′ positions; polymerisation proceeds 5′→3′ via phosphodiester formation, a geometry enforced by the sugar ring and leaving groups. Each phosphate carries a negative charge at physiological pH, imparting a uniform polyanionic character to the backbone. Far from being an inconvenience, this uniform charge is precisely what keeps long chains soluble, discourages nonspecific collapse, and channels recognition towards the bases through predictable hydrogen bonding and stacking rather than through backbone–backbone adhesion (Watson and Crick 1953; Saenger 1984; Westheimer 1987). In water, where solvent competition and ions can frustrate weak interactions, this architecture makes nucleic acids readable and copyable: the backbone is the regular rail; the bases are the legible characters; base pairing is the rulebook for templating.

Taken together, a clear division of labour emerges. Carbon supplies stable bonds, abundant architectures, and the possibility of long chains. Amide chemistry folds those chains into machines that act. Phosphate–sugar backbones provide polarised, uniformly charged rails on which digital-like messages can be written, stored, and copied with high fidelity in water. Once a backbone can be built in one direction, remain soluble as it grows, and present bases for specific pairing, heredity can be both robust and evolvable. Chemistry furnishes the grammar; sequences can then encode, persist, and be acted upon.

From this general chemical logic, the move to plausible origins settings is natural rather than abrupt. Ordinary geologic environments supply the stage and many of the props. On volcanic coasts and ash-rich basins, thin hydrated films of weathered glass and expandable clays such as montmorillonite both concentrate small organics and catalyse phosphodiester formation from activated ribonucleotides, yielding short RNAs with reproducible sequence biases. A series of classic studies by Ferris and co-workers showed that layered silicates not only accelerate RNA chain growth but also guide oligomerisation into stretches tens of nucleotides long in aqueous suspensions, without enzymes (Ferris and Ertem 1993; Ferris 2002). These films act as concentrators and kinetic guides, converting an unstructured mix into an enriched subset of sequences—the kind of asymmetry that later compartments and cycles can amplify.

Compartmentalisation arrives with equal simplicity. Short single-chain fatty acids—plausibly supplied by Fischer–Tropsch-type catalysis or hydrothermal synthesis—spontaneously assemble into vesicles upon hydration, grow osmotically when fed amphiphiles, and divide under shear or filament tension. Numerous experiments in unremarkable aqueous conditions show vesicles permeable to small nutrients yet retentive of longer polymers (Hanczyc, Fujikawa, and Szostak 2003; Mansy and Szostak 2008; Budin and Szostak 2011). When mineral films holding short RNAs are wetted, co-encapsulation follows: some vesicles trap RNAs together with carboxylates, metal ions, and buffering anions from the same microenvironment. Heterogeneity across vesicles—often dismissed as "messy"—here provides the variation on which protocell-level selection can act.

Within such microreactors, template-directed copying no longer rests on conjecture. Activation with 2-aminoimidazole generates imidazolium-bridged dinucleotide intermediates that account for the kinetics and sequence preferences of non-enzymatic primer extension on RNA. These pathways add multiple nucleotides at rates meaningful on geologic timescales and have been characterised structurally and mechanistically (Li et al. 2017; Walton et al. 2020; Zhang et al. 2022). Environmental cycling helps rather than hinders: freeze–thaw and wet–dry pulses—common in littoral and geothermal settings—can both activate and extend in the same mixture, mirroring natural rhythms of hydration and temperature (He et al. 2017; Becker et al. 2019). A classic incompatibility—RNA copying benefits from Mg^{2+}, yet Mg^{2+} destabilises fatty-acid membranes and promotes RNA cleavage—can be tuned chemically. Citrate, a small tricarboxylate that is prebiotically plausible via meteoritic delivery and non-enzymatic carboxylate networks, chelates Mg^{2+} sufficiently to protect both membranes and RNA while leaving enough free Mg^{2+} to sustain templated synthesis; with citrate present, primer extension proceeds inside fatty-acid protocells (Adamala and Szostak 2013). The vesicle thus shifts from container to reactor.

Peptides plausibly arise alongside these processes and assist long before ribosomes exist. Wet–dry cycling of α-hydroxy acids with amino acids yields depsipeptides that rearrange into peptide-rich mixtures upon subsequent cycles; among the products, short cationic amphiphiles stabilise fatty-acid membranes and bind RNA, and mixed backbones can facilitate further reactivity (Forsythe et al. 2015; Frenkel-Pinter et al. 2021). In parallel, in vitro selection has produced ribozymes that aminoacylate substrates and even catalyse primitive peptidyl-transfer reactions, indicating that RNA alone can perform translation-like chemistry; the modern ribosome's catalytic heart—the peptidyl transferase centre—is RNA, consistent with this evolutionary layering (Saito, Kourouklis, and Suga 2001; Turk, Chumachenko, and Yarus 2010; Nissen et al. 2000). Together, these results point to an early RNA–peptide mutualism: peptides stabilise membranes and assist RNA folding; RNAs, in turn, catalyse reactions that yield more—and better—peptides.

Once inside vesicles, macromolecular crowding favours RNA folding and ligation, and encapsulated catalysts can measurably advantage their host compartments in competition experiments. Vesicles that accumulate more polymers swell osmotically, recruit amphiphiles faster, and cross division thresholds more often; small gains in copying length or stability compound at the protocell level, enriching chemistries that extend and preserve longer RNAs (Mann 2012; Niederholtmeyer et al. 2015). Laboratory evolution has also pushed RNA polymerase ribozymes past nontrivial milestones: ribozyme-catalysed transcription of another active ribozyme has been demonstrated (Wochner et al. 2011), and sustained amplification by an RNA polymerase ribozyme has been achieved (Horning and Joyce

2016). Self-sustained replication thus becomes a matter of degree, not a binary leap.

At this juncture, a transparent pivot is important because what follows connects these established waypoints to a forward-looking proposal. The evidence just reviewed—surface-guided RNA oligomerisation, fatty-acid vesicle growth and division, citrate-enabled copying within protocells, non-enzymatic and ribozyme-mediated polymerisation, and peptide assistance—defines a plausible corridor. It does not, by itself, specify a single historical route. The next paragraphs present a coherent hypothesis for how DNA might have been recruited as a more durable archive within protocells. Each component is anchored in reported chemistry or biophysics, but the sequence of events—especially an incremental drift from RNA templating to RNA→DNA templating inside vesicles—remains unproven. Readers should treat it as a testable scenario, not a reconstruction of fact.

With that framing, consider three ingredients: availability of precursors, catalytic promiscuity, and selection at the protocell level. Availability is supported by multiple routes. Meteoritic influx and photoredox networks feed sugars and nucleobases; ribose has been identified in carbonaceous meteorites with extraterrestrial isotopic signatures (Furukawa et al. 2019), and non-enzymatic protometabolic cycles driven by metals can generate tricarboxylic-acid intermediates from simple inputs such as glyoxylate and pyruvate (Muchowska, Varma, and Moran 2019). Photoredox chemistry can, under overlapping conditions, produce both ribo- and deoxy-nucleosides, implying that mixed nucleotide pools may have existed in the same locales (Patel et al. 2015; Xu et al. 2020). Catalytic promiscuity follows from how ribozymes evolve: laboratory-selected RNA polymerase ribozymes are not rigid protein enzymes; their active sites tolerate mutation and structural variation as processivity improves, making occasional accommodation of a deoxynucleotide among ribonucleotides chemically credible (Horning and Joyce 2016). Selection then acts through vesicle ecology. In citrate-compatible protocells where non-enzymatic and ribozyme-assisted copying already occur, any event that yields a slightly more hydrolysis-resistant template should persist longer and be recopied more often over environmental cycles.

This book advances a testable hypothesis. In niches where ribonucleoside and deoxyribonucleoside triphosphates coexist, an evolving RNA polymerase ribozyme—or a small RNP—could occasionally accept a deoxynucleotide during RNA-templated synthesis, creating short, more hydrolysis-resistant cDNA segments. Most such accidents would vanish with their vesicles; a minority might persist long enough to be recopied across environmental cycles, conferring a slight protocell-level advantage. Laboratory ribozymes have already shown that RNA enzymes can accept deoxynucleotides and copy RNA into DNA, establishing biochemical feasibility for occasional dNTP acceptance by small RNP catalysts (Samanta and Joyce 2017; Tjhung et al. 2020; Horning and Joyce 2016; Attwater et al. 2018). Because vesicle growth and reproductive success correlate with internal polymer load and the retention of functional templates, protocells that, by chance, accumulate a somewhat more "DNA-like" archive would tend to divide more often and seed subsequent generations more successfully. Over many cycles, that small bias could drift the catalytic population towards weak RNA→DNA activity—the conceptual ancestor of reverse transcriptase—not through foresight, but by favouring what best preserves heritable information (Joyce and Orgel 1999; Benner, Kim, and Yang 2012; Hudson et al. 2013; Adamala and Szostak 2013). We emphasise that this remains to be demonstrated as a continuous, intravesicular pathway; it synthesises established waypoints into an experimentally approachable scenario and invites tests in microfluidic landscapes and cyclic geochemical reactors that alternate feed, dehydration, rehydration, and shear.

Because provenance matters, it is stated plainly that the vesicle-hosted drift from RNA copying to RNA→DNA templating described earlier is presented here as an original, evidence-informed hypothesis rather than as an established historical account. The mechanistic steps have experimental anchors—mineral-assisted oligomerisation (Ferris and Ertem 1993; Ferris 2002), fatty-acid vesicle growth and division (Hanczyc et al. 2003; Budin and Szostak 2010), citrate-tuned Mg^{2+} chemistry enabling intravesicular copying (Adamala and Szostak 2013), 2-aminoimidazole activation and imidazolium-bridged intermediates (Li et al. 2017; Walton et al. 2020; Zhang, Szostak, and Li 2022), ribozyme-mediated replication gains (Wochner et al. 2011; Horning and Joyce 2016), and convergent photoredox routes to both ribo- and deoxy-nucleosides (Patel et al. 2015; Xu et al. 2020), but the particular sequence and population-level ratchet remain to be demonstrated end-to-end in continuous laboratory systems. Framed this way, the scenario is not a claim of historical certainty; it is a blueprint for experiments.

Returning to the didactic arc, the core chemical logic can be restated without repetition. Carbon's covalent toolkit enables polymers with controllable geometry and aqueous stability. Amide linkages transform linear strings into machines by constraining local geometry and enabling reproducible secondary motifs whose side chains populate catalytic landscapes. Phosphate–sugar backbones add the final elements required for digital-like information: uniform charge, directionality, and regularity that favour templated copying while maintaining solubility. When such polymers are supplied with plausible activation chemistries and housed in simple amphiphile compartments that grow and divide, selection can act on the properties of the polymers at the protocell level. No leap of faith is required to arrive at a system in which heritable information appears, influences its own persistence, and is progressively refined.

It helps to return to first principles. Carbon's uniqueness matters because it alone reliably provides strong, directional bonds for long backbones that can be heavily functionalised without losing integrity, while simultaneously tolerating polar, charged, and aromatic groups that tune solubility and recognition in water. From there, amide planarity turns strings into shapes that do work, and the phosphate–sugar backbone turns strings into addresses that can be read and rewritten. The remainder—compartments, catalysts, cycles, and, possibly, the opportunistic recruitment of DNA as a tougher archive—follows from chemistry in context, refined by selection wherever persistence and reproduction become possible. With that material grammar in place, the next question is temporal: How does information written into these substrates shift its weight across generations?

In evolutionary theory, the same logic acquires temporal depth. Population genetics treats allele and genotype frequencies as state variables whose dynamics follow from the joint action of selection, mutation, recombination, and drift (Fisher 1930; Wright 1931; Kimura 1968; Crow and Kimura 1970). The "deterministic spine" corresponds to selection and inheritance—constraints that canalise what endures—an idea that Waddington made vivid as canalisation and the epigenetic landscape, now naturally reframed in informational terms as constraint architectures shaping flow over phenotype space (Waddington 1942). Mutation and recombination supply variance that explores new regions of genotype space; regulation and gene networks reweight mappings from genotype to phenotype across environmental contexts (Jacob and Monod 1961; Wagner 2014); developmental and ecological plasticity alter those mappings in real-time (West-Eberhard 2003). In finite populations, demographic stochasticity and sampling error—drift—ensure that even under identical rules, replicate populations trace distinct but lawful frequency trajectories, a fact captured by Wright–Fisher and Moran processes and their diffusion limits (Ewens 2004). At a higher level of abstraction, the Price equation makes transparent how changes in trait means decompose into covariance (selection) and transmission terms (inheritance, mutation, learning), again emphasising distributions and information flow rather than single paths (Price 1970).

On this view, indeterminism is not an exception to the lawful order. It is a property of our effective descriptions of systems with many degrees of freedom—rooted in missing variables, fluctuating inputs, finite measurement, and changing boundary conditions—not a breakdown of the underlying equations. Put differently, the update rule can remain fixed while the distribution it propagates depends on what information is resolved, what is ignored, and how the environment couples into the system (Khinchin 1949; Jaynes 1957). In physics this appears as ensemble dynamics and coarse-grained equations for probability densities; in evolution it appears as replicator–mutator and diffusion approximations for allele distributions, where selection supplies directional constraint and stochastic terms capture unresolved microprocesses and demographic noise (Kimura 1964; Crow and Kimura 1970; Ewens 2004). Laws set the admissible region and the conservation structure; information flow—inputs, couplings, controls, and inherited code—decides which lawful path is actually taken through that region. In this sense, prediction in evolutionary systems is distributional by nature: we forecast ensembles of possible, law-respecting futures and track how evidence and constraint shift probability mass among them, rather than expect a single fate from a single rule (Jaynes 1957; Fisher 1930; Wright 1931).

In short, distributions capture constraints and opportunities without implying purpose.

Crucially, this view is not teleological: life doesn't merely adapt; under lawful dynamics and information constraints, it "discovers."

This chapter proposes a metaphorical reframing: evolution as a feedback-driven information system. This reframing draws on foundational work in cybernetics (Wiener 1948) and systems biology (Noble 2006), where feedback loops are central: changes in the environment provoke changes in organisms, which in turn alter the environment—a recursive, dynamic interplay.

This perspective aligns with the Modern Synthesis but extends it by recognising that the adaptive process is not a one-way street; it is a dialogue between genes and environment (Lewontin 2000), a continual recalibration towards greater fitness.

Natural selection, when viewed through the lens of systems science and information theory, becomes a distributed, parallel search engine—fine-tuning biological structures to meet constraints dictated by physics and chemistry. Genes are not merely carriers of traits; they function as informational hypotheses—not consciously, but as provisional blueprints whose value is tested against the world, one generation at a time. Each gene variant, or allele, is a "proposal" about what might work in a given context, retained or discarded depending on its success in the shifting ecological theatre. This concept, sometimes called "evolutionary epistemology," is robustly argued by Dawkins in *The Extended Phenotype* (1982) and expanded in the context of kinship by Haig (2002).

Importantly, this view is not teleological: there is no foresight or purpose, only the accumulation of "good guesses" that have survived the culling power of selection (Dennett 1995).

And the patterns repeat.

From flight to echolocation, from photosynthesis to quantum tunnelling in enzymes, the solutions of life echo the very principles that scientists unravel with computation and experimentation—a view supported by research (Ball 2011), a feature article that surveys emerging, still-debated examples. It's as if biological evolution had long been running a natural form of biophysical reverse engineering, hacking into the source code of the universe to extract functional elegance—a metaphor explored in Kauffman (1993) and Conway Morris (2003), who argue that evolution explores the space of possible forms, arriving, again and again, at functionally inevitable solutions.

In this light, organisms are not mere survivors of environmental pressure. They are embodied solutions to complex differential equations, evolved not in silico but in vivo—an analogy advanced by Wolfram (2002), who explores how simple computational rules can generate complex biological patterns. The eagle's wing, the dolphin's echolocation, the spider's web—all arise from a silent yet profound dialogue between living form and the invariant laws of the universe. As Richard Lewontin observes in *The Triple Helix* (2000), organism and environment are not separate actors but co-construct one another over evolutionary time. Each adaptation is thus both a response to external constraint and a creative act within an ever-evolving context—life and world entangled in a continual process of mutual shaping.

And so, we arrive at a deeper question: Did the eagle's anatomy reflect an intuitive grasp of fluid dynamics, or was it shaped by the very laws that govern air and motion? In truth, it was not intuition, but evolution—acting over deep time—that sculpted biological structures in response to aerodynamic constraints. Through repeated interactions with the environment, form became optimised for function. The principles that aerospace engineers would later derive in equations and wind tunnels had already been inscribed into feather and bone—not by conscious design, but by the persistent filtering of natural selection acting on physical law.

And perhaps our own discovery of these laws is not invention, but recognition—a belated decoding of the informational fabric that life has been writing all along. As Michael Polanyi observes in *Personal Knowledge*, scientific discovery often consists in uncovering realities that pre-exist our awareness of them.

This chapter begins with a provocation: What if evolution is more than random variation filtered by survival? What if it is a form of emergent engineering—a distributed, information-driven algorithm capable of "solving" the principles of aerodynamics, not by theory, but by embodiment?

As engineers reverse-engineer the bird, they find their blueprints already drawn—in bone, tendon, and plume. Evolution, it seems, didn't just stumble onto flight. It optimised it—iteratively, informatically—using the sky as its open laboratory and the winds as feedback.

This remarkable capacity for refinement is at the heart of evolution, a foundational concept in biology that explains how species adapt over time to their environments. Traditionally, evolutionary theory, as outlined by Charles Darwin (1859), has focused on the gradual process of genetic mutation and natural selection, where random changes in genes lead to advantageous traits that improve survival and reproduction.

Conventional evolutionary theory is sometimes portrayed as treating DNA as a passive storehouse acted upon by random mutation, but since the 1960s, biology has emphasised active, environment-responsive gene regulation. Building on that foundation, research on epigenetic inheritance and predictive cellular responses has reinforced a more dynamic picture of heredity—one in which regulatory networks and chromatin states help organisms tune expression to context, with some effects persisting across generations (Jacob and Monod 1961; Jablonka and Lamb 2005)

In this emerging framework, genetic material functions more like a dynamic system of information management, constantly interacting with internal and external stimuli, in line with theories described by Shapiro (2011).

Think of it this way: every organism exists at the boundary between order and chaos. Its genome is not just a blueprint; it's an adaptive interface—a real-time operating system that listens to its surroundings. When food becomes scarce, or a predator's scent lingers in the wind, organisms don't merely wait for mutations to provide a solution. They adjust, epigenetically modulating gene expression patterns to survive the moment and sometimes even transmit those adaptive settings to the next generation. This is evolution in conversation with the environment, not a monologue of randomness.

At the molecular level, this means that transcription factors, enhancer regions, and feedback loops allow for a vast spectrum of gene expression, depending on the context. A gene is no longer a static piece of code but a conditional algorithm—executing different outputs based on different environmental "if-then" clauses.

In this light, natural selection itself becomes a mechanism for information refinement. It's not simply about traits that improve survival; it's about improving the fidelity and functionality of how information is encoded, accessed, and applied. Evolution starts to resemble a learning process, where successful strategies—stored as DNA sequences, regulatory motifs, and expression patterns—are archived and propagated. Failed experiments are forgotten. Over time, this leads to genomes that are not only fit for survival but fit for understanding—better attuned to the structural and behavioural demands of their environment.

This also aligns with theories from information thermodynamics and complex systems science, where adaptation emerges not from chaos but from feedback-rich environments that reward information coherence.

2.2 Genes as Hypotheses: Evolution as an Epistemic Feedback Loop

The genome, in this context, is not a script—it is an evolving dialogue, a recursive structure through which evolution reveals itself not merely as the accumulation of biological changes but as a deeper process—a progressive optimisation of how life absorbs, encodes, and deploys information, a concept powerfully advanced by Gell-Mann (1994) and by Noble (2006). Once thought to be static records of the past, genes emerge instead as dynamic hypotheses—continuously tested against the shifting canvas of the environment and refined through generational feedback loops. Of course, this "hypothesis" is not a conscious conjecture, but a metaphor drawn from evolutionary epistemology: each genetic variant is proposed to the world, its value judged by survival. This epistemic view of genes as evolving predictions about the world is argued by Richard Dawkins (1982) and resonates with David Haig (2002).

And here lies the deeper insight—perhaps the most profound of all: Life doesn't simply adapt to information; life is information—persisting, improving, and remembering.

By interpreting evolution as the continual optimisation of information flow, we move beyond the classical notion of adaptation as mere physical change. Instead, adaptation becomes the iterative refinement of how organisms "read," respond to, and anticipate their environments—a perspective rooted in Shannon (1948), which quantified information as the reduction of uncertainty, and philosophically advanced by Dennett (1995), who argued that natural selection functions as an algorithmic filter, preserving the most reliable, context-sensitive responses over time. Adaptation, from this vantage, is about the honing of informational congruence between organism and environment, rather than any teleological goal. Seen through this lens, evolution is not simply the selection of physical traits, but the recursive refinement of informational structures. By framing evolution as the optimisation of information flow, we open a richer, more coherent understanding of life's unfolding. This is a view supported in Adami (2016) and echoed by Floridi (2011).

"Information" here refers not to Shannon entropy alone, but to structured distinctions that reduce uncertainty within an adaptive system. "Optimisation" refers to iterative refinements in phenotype-environment alignment, not teleological improvement. This non-teleological notion of optimisation is central to contemporary systems biology and evolutionary computation.

Adaptation is no longer about merely selecting physical traits. It becomes a recursive honing of organisms' capacities to detect, interpret, store, and act upon the environmental signals that shape their fate. This concept finds support in West-Eberhard (2003).

In human society, the concept of evolution is often tied to improvements in how we gather, process, and share information. Throughout history, advancements from the printing press to the internet have reflected a societal drive to improve information flow, a core aspect of progress and adaptability. This analogy is thoughtfully explored by Gleick (2011), who traces the history of information as both a physical and biological phenomenon. Just as these technological advancements have reshaped human society, biological systems can be seen as refining ways to "store" and "process" environmental information to increase their chances of survival and reproductive success—echoes in and by Davies (2019), who argues that information is a causal agent in both life and technology. This view implies that evolution is not only about physical changes but also about improving the efficiency of encoding, retrieving, and acting on information, suggesting a dynamic, adaptive network continuously honing its ability to interact with and adapt to its surroundings.

In this context, genes are not simply carriers of physical traits but repositories of biological data that has been selected for its relevance to survival. This position finds support in Maynard Smith (2000). Genetic mutations can thus be seen as modifications to information encoding rather than arbitrary alterations in a species' biology. When a mutation occurs, it represents a shift in the genetic code that might change how effectively an organism can interpret and respond to environmental cues. In this view, natural selection functions as a sophisticated filtering mechanism, favouring mutations that enhance the efficiency and relevance of genetic information in ways that align with the environmental conditions. Rather than simply selecting for advantageous physical traits, natural selection refines the organism's information processing capabilities, prioritising mutations that allow genes to better "tune in" to the specific demands of their surroundings.

In this light, the adaptive filtering performed by natural selection is not a static event, but an ongoing, iterative dialogue between organism and environment. Genetic information is not passively inherited but is continually reshaped in response to the constant flow of external "data." Environmental factors, therefore, do not simply impose constraints; they act as active participants in this exchange, exerting selective pressures that guide and sculpt the genetic blueprint. This concept echoes in the research paper "Phenotypic Plasticity and Epigenetic Marking: An Assessment of evidence for genetic accommodation," *Evolution*, 2014, by Schlichting, C.D. and Wund, M.A.

Through countless generations, this ceaseless interaction leads to an ever-finer alignment between genetic instruction and ecological reality. The result is a lineage whose capacity to sense, interpret, and respond to its environment becomes increasingly sophisticated. The information stored within the genome is not only preserved for survival's sake but is honed and recalibrated, generation after generation, until it mirrors with greater fidelity the shifting demands of the world outside. This interpretation of natural selection brings a dynamic quality to evolution, framing it as a process where information flows between organisms and their environments in a mutual feedback loop. As a result, life's adaptability becomes a matter of evolving informational congruence, where genetic changes are not just random events but refinements that actively improve how well an organism "reads" and interacts with its environment.

Furthermore, cells do not merely hold genetic information passively; they actively regulate and express genes in response to external signals. The regulatory network of a cell operates as a dynamic system, constantly encoding and transmitting real-time information to ensure that the organism responds appropriately to changing conditions. This enables cells to optimise energy usage and function efficiently, reflecting a form of information processing that is highly adaptable to immediate demands. The complexity of regulatory networks, which allow for flexible gene expression in response to shifting external conditions, is beautifully demonstrated in Eric Davidson's *Genomic Regulatory Systems* (2001).

As organisms become more complex, their ability to process environmental information and respond to stimuli becomes increasingly sophisticated. Nervous systems, particularly in animals with more advanced cognitive functions, represent a significant leap in the evolution of information processing (Seth 2016). The transition from simple reflexive responses to complex memory, decision-making, and abstract thinking demonstrates an evolutionary drive towards intricate forms of data processing. Nervous systems are not simply survival tools; they serve as advanced processors enabling organisms to model their environments, anticipate potential outcomes, and make informed decisions. These capabilities are an evolutionary advantage, as they allow organisms to adapt rapidly and flexibly to both internal and external changes.

In highly social species, evolution has produced an additional layer of information flow optimisation. The emergence of complex social hierarchies, communication systems, and cumulative cultural learning has enabled these organisms to construct external networks of shared information. This phenomenon is explored in depth by Dunbar, R.I.M., *Grooming, Gossip, and the Evolution of Language* (1996), who demonstrates that the size and complexity of primate social groups correlate with neocortical volume and the richness of social cognition. Here, evolution selects not only for genetic adaptation, but for the capacity to transmit, process, and build upon information in a social context, accelerating behavioural innovation and collective problem-solving.

In social animals, survival strategies, knowledge, and even cultural practices are passed from one generation to the next, allowing for a collective evolution that surpasses the individual's genetic information. Humans, in particular, have developed languages, arts, and technologies that create shared repositories of knowledge, extending information networks beyond individual organisms. Human society, therefore, can be seen as an elaborate informational ecosystem where survival and progress are increasingly tied to the optimisation of collective data storage and retrieval.

At the scale of ecosystems, biological communities function as intricate networks of information exchange, where every interaction—predation, competition, symbiosis, or resource cycling—represents a transfer or transformation of environmental data. This systems-level understanding of ecology is championed by Levin, S.A., "Ecosystems and the Biosphere as Complex Adaptive Systems," *Ecosystems*, 1998, who argues that stability, resilience, and adaptability emerge from the distributed, feedback-driven information flows that connect species within the biosphere. Such a framework allows us to see the ecosystem not as a mere collection of organisms, but as an evolving "information architecture" that continuously adapts to perturbations and opportunities. Predator–prey relationships, symbiotic partnerships, and competitive interactions create dynamic webs of information transfer, where each species adapts to others in a co-evolutionary process that supports survival within the ecosystem. This information-centric view suggests that ecosystems are not simply collections of organisms but highly structured networks that rely on the optimal flow of information for stability and resilience. Evolution, in this light, becomes a collaborative optimisation of information channels where each species plays a role in maintaining the ecosystem's "information architecture."

Viewing evolution as an optimisation of information flow provides fresh insights for scientific research across fields. In neurobiology, this perspective could inspire new approaches to understanding the brain not merely as a tool for survival but as an optimised system for decoding, encoding, and transmitting information. This framework might offer novel explanations for complex phenomena like memory formation, consciousness, and learning. In ecosystem science, perceiving ecosystems as interconnected information networks could lead to strategies that focus on conserving species not only for biodiversity's sake but also for their unique contributions to the ecosystem's data flow. Such an approach could redefine how we understand and manage ecosystems, prioritising species and interactions that support the network's stability and adaptability.

This perspective on evolution does not replace the classical genetic model but instead complements it. Natural selection remains central to evolution, but the focus expands to include mechanisms that improve an organism's informational capacities. Evolution, from this viewpoint, becomes a process continuously refining life's ability to process and act upon information, evolving towards systems of greater complexity and resilience in a world of constant change, a perspective championed in Heylighen, F., "Complexity and Information: A New Evolutionary Paradigm," *Futures*, 1999. Evolution, then, is the unfolding of a vast adaptive network where genes, neural circuits, social customs, and communication networks work together to encode, decode, and respond to the world's data, ensuring survival through optimised information flow.

2.3 Epigenetics: How Genes Gather Information to Adapt

In 1859, Charles Darwin revolutionised scientific thought by publishing his groundbreaking work, *On the Origin of Species*, which introduced the concept that species evolve over time through natural selection. According to Darwin, individuals within a species vary in their traits, and these variations are largely caused by random genetic mutations. In any given environment, some traits provide a survival advantage, allowing individuals with those traits to reproduce more successfully. Over time, these traits become more common within the population, leading to the evolution of the species.

Darwin's theory of natural selection rests on three core principles: variation, competition, and inheritance. Variation refers to the differences in traits among individuals of a species, such as size, colour, or behaviour. Competition arises due to limited resources like food, water, and shelter, which means that not all individuals can survive and reproduce. Those individuals with traits that provide a competitive edge are more likely to survive and pass those traits on to their offspring. Inheritance ensures that these advantageous traits become more common over generations, driving the evolution of the species.

One of the key strengths of Darwin's theory is its simplicity. It provides a straightforward explanation for how species evolve: through the interplay of variation, selection, and inheritance. However, Darwin did not understand the genetic mechanisms underlying inheritance. Unbeknownst to many at the time, Gregor Mendel's work set the groundwork for modern genetics. Thus, Darwin's theory lacked a detailed explanation of how traits passed from one generation to the next.

In the early 20th century, by synthesising Darwin's theory of natural selection with Mendelian genetics, biologists created what is now known as the Modern Synthesis. This comprehensive framework provided a deeper understanding of how evolution operates by integrating natural selection with the principles of heredity. The Modern Synthesis showed that mutations in DNA create genetic variability, and natural selection chooses which traits get passed on to the next generation by working on this variability. This foundational framework is best represented in classic works such as *Genetics and the Origin of Species* by Theodosius Dobzhansky (1937) and *Evolution: The Modern Synthesis* by Julian Huxley (1942).

According to this view, by introducing new traits into a population, evolution is propelled by random mutations. If a mutation provides a survival or reproductive advantage, individuals carrying that mutation are more likely to reproduce and pass the trait to their offspring. Over time, natural selection weeds out deleterious traits, making advantageous traits more common within the population.

The Modern Synthesis reinforced Darwin's ideas with a genetic foundation but retained the basic premise that random mutations are the primary source of variation and that natural selection is the primary driver of evolutionary change. However, as our understanding of biology has grown, it has become clear that the Modern Synthesis cannot explain all aspects of evolution. For instance, explaining the rapid adaptability of certain species, the apparent non-randomness of some evolutionary changes, and the role of environmental factors in shaping gene expression is a challenge. This has led to the development of new theories, such as epigenetics, which provide a more nuanced understanding of how species evolve. A seminal critique of these limitations is found in Mary Jane West-Eberhard's *Developmental Plasticity and Evolution* (2003).

Epigenetics refers to heritable changes in gene expression that do not alter the DNA sequence itself but can nonetheless be triggered by environmental cues—through mechanisms like DNA methylation, histone modification, and non-coding RNAs. The field has been transformed by landmark work such as Jirtle and Skinner's (2007) review of environmental epigenomics, and Heijmans et al.'s (2008) work on the Dutch Famine, which demonstrated persistent, environmentally induced methylation differences in human populations.

It is important to clarify: while such epigenetic marks can sometimes persist across generations (e.g., in plants: Boyko et al. 2010), the evidence for widespread, stable transgenerational epigenetic inheritance in mammals remains limited and is the subject of ongoing debate (Heard and Martienssen 2014). One of the key insights from epigenetics is that environmental factors, such as diet, stress, toxins, and social interactions, can influence gene expression. These changes can have lasting within-lifetime effects and sometimes intergenerational consequences; in mammals, strong transgenerational inheritance is limited and remains debated. Jirtle and Skinner (2007) offer a comprehensive review of environmental epigenomics and inheritance in mammals. This phenomenon, known as transgenerational epigenetic inheritance, challenges the traditional notion that random mutations are the sole drivers of evolutionary change. While there is growing evidence for transgenerational epigenetic effects, most biologists caution that these mechanisms supplement, rather than supplant, the established role of DNA sequence mutations in long-term evolutionary change.

To comprehend how epigenetics changes the evolutionary paradigm, one must investigate the mechanisms through which epigenetic modifications take place. There are several known mechanisms by which epigenetic changes can occur, each of which plays a role in regulating gene expression.

a) DNA Methylation: One of the most well-understood mechanisms of epigenetic regulation is DNA methylation, which involves the addition of methyl groups to specific regions of the DNA molecule, typically at cytosine bases in CpG islands. Methylation often silences genes, preventing them from being transcribed into RNA and subsequently translated into proteins. This form of gene regulation is crucial during development, allowing cells to differentiate by activating or silencing specific genes.

b) Histone Modification: In eukaryotic cells, DNA is wrapped around proteins called histones, forming a structure known as chromatin. Modifications to these histones, such as the addition of acetyl or methyl groups, can alter how tightly or loosely the DNA is wound around them. When tightly wound, DNA becomes inaccessible to the transcription machinery, effectively silencing the associated genes. Conversely, when DNA is loosely wound, the genes are accessible for transcription, allowing them to be expressed. Histone modifications provide a dynamic and reversible way to regulate gene expression in response to environmental cues.

c) Non-coding RNA Molecules: Non-coding RNAs (ncRNAs) are small RNA molecules that do not code for proteins but play a crucial role in regulating gene expression. ncRNAs can bind to messenger RNA (mRNA) and prevent it from being translated into protein, or they can interfere with the transcription process by interacting with DNA or proteins involved in gene regulation. This adds another layer of complexity to the regulation of gene expression, allowing cells to fine-tune their response to environmental stimuli.

These mechanisms provide a flexible and dynamic system for regulating gene expression in response to environmental changes. While genetic mutations are typically random and permanent, epigenetic modifications can be triggered by external factors and are frequently reversible. This allows organisms to quickly adapt to changes in their environment without altering their underlying DNA sequence.

Mutations are random and irreversible in traditional Darwinian evolution. Epigenetics introduces a more flexible and responsive mechanism of adaptation. It suggests that organisms can rapidly respond to environmental changes through epigenetic modifications, which can influence the evolution of populations. This adaptability introduces a new layer of complexity to evolutionary theory by allowing organisms to "sense" their environment and adjust their genetic expression in real-time, without waiting for random mutations to occur.

Studies show that severe prenatal famine is associated with durable epigenetic marks into adulthood in exposed individuals (intergenerational effects); robust, stable transmission to unexposed later generations in humans remains uncertain and debated. A well-known cohort study found that individuals exposed in utero during the Dutch Hunger Winter showed lasting DNA methylation differences (e.g., IGF2) into adulthood—an intergenerational effect in the exposed cohort; reliable transmission to unexposed descendants has not been demonstrated in humans. The Dutch Hunger Winter (1944–1945) shows that severe prenatal famine leaves durable marks in people exposed in utero. Adults conceived during the famine display lasting epigenetic and phenotypic differences—famously, ~5% lower DNA methylation at the IGF2 DMR decades later—with timing-specific changes also reported at loci such as LEP, ABCA1, GNASAS, and IL10, alongside metabolic/cardiovascular associations. These are intergenerational effects; robust transgenerational inheritance in humans—that is, in generations not directly exposed—has not been demonstrated.

The ability of organisms to adjust their gene expression in response to environmental changes offers a more dynamic view of evolution. Epigenetics proposes that organisms actively participate in their evolution by constantly adapting their gene expression to enhance survival and reproduction, rather than merely undergoing a slow process of random mutations.

Epigenetics advocates for the idea that information is a key driver of evolution. It suggests that genes, far from being passive repositories of information, are actively "listening" to their environment and adapting in real-time to ensure survival. Epigenetic regulation shows that gene expression is highly responsive to environmental cues. While such changes can shape phenotypes and sometimes persist to the next generation, long-term evolutionary change still primarily tracks DNA sequence variation. This concept aligns with the idea that information is a living, evolving force—a view that extends beyond biology into the realm of systems biology, where life and the universe are interconnected by the flow of information.

The idea of information as a central force in evolution is not entirely new. Claude Shannon's theory of information, which emerged in the mid-20th century, revolutionised how we think about communication and data transmission. Shannon's theory provided a mathematical framework for quantifying information and demonstrated how information could be processed, transmitted, and decoded. In the biological context, by incorporating the signals received from their environment and how these signals are processed by their genetic and epigenetic systems, the concept of information can be broadened.

The role of information in evolution is clearest when organisms must decide whether a change is momentary noise or a meaningful shift that warrants a lasting adjustment. Epigenetic regulation helps translate sustained signals—such as prolonged scarcity, repeated stress, or persistent temperature change—into coordinated changes in gene expression. In this way, the organism does not merely react to the present; it tunes itself toward the world it expects to endure.

Studies have shown that when plants experience drought conditions, undergoing epigenetic changes can help them conserve water. These changes include the activation of genes involved in water transport and the suppression of genes that promote water loss. This ability to gather and process environmental information allows plants to survive in harsh conditions and pass on these adaptive traits to their offspring.

In social animals, some of the strongest evolutionary pressures arise not from climate or hunger, but from social interaction. Social interactions, for instance, can influence gene expression in ways that affect behaviour, reproduction, and survival. For example, studies have shown that exposure to social stress can lead to epigenetic changes in brain function, which in turn affects behaviour. In some cases, these changes can be passed on to future generations, suggesting that social information can shape evolutionary outcomes.

So how does the environment 'speak' to the genome without changing its sequence? The term "epigenetics" comes from the Greek word "epi," meaning "above" or "on top of" and refers to regulatory mechanisms that control gene expression without altering the underlying genetic code.

Traditional Darwinian evolution assumes that genetic mutations occur randomly, and those that confer a survival advantage are selected for over generations. Epigenetics, however, introduces the idea that environmental factors can directly influence gene expression, leading to rapid, reversible adaptations that do not require changes in the underlying DNA.

a. Lamarck Revisited: In the early 19th century, Jean-Baptiste Lamarck proposed that organisms could pass on traits acquired during their lifetimes to their offspring. Though Lamarck's theory was largely dismissed, epigenetics shows a narrow sense in which acquired-like states can sometimes be inherited in specific systems; this does not restore Lamarckian inheritance as a general mechanism. Environmental factors can indeed influence gene expression, and these changes can be inherited, challenging the traditional view that only genetic mutations are heritable. However, such environmentally induced epigenetic states typically persist only for a generation or two in specific systems and do not restore Lamarckian inheritance as a general mechanism of evolution.
b. Adaptive Flexibility: Epigenetics allows for greater adaptive flexibility than random mutations. While mutations occur randomly and often slowly, epigenetic changes can happen rapidly in response to environmental challenges. Plants facing drought, for example, can undergo epigenetic changes that enhance their resilience to drought, and these characteristics can be passed down to their descendants. This provides a more immediate mechanism for adaptation than traditional genetic evolution.
c. Environmental Information: The ability of environmental factors to directly influence gene expression supports the idea that environmental information is a driving force in evolution. Genes are not isolated entities but are constantly interacting with their surroundings, "listening"

to environmental cues and adjusting in real-time to ensure survival. This aligns with the concept that information is a living, evolving force shaping life.

Case Studies in Epigenetics and Evolution

1. The Dutch Famine Study
   A widely cited example of intergenerational epigenetic effects in humans is the Dutch Famine of 1944–1945, also known as the Hunger Winter. During this period, severe food shortages in the Netherlands resulted in widespread malnutrition, particularly affecting pregnant women and their developing children. Decades later, research revealed that individuals who were exposed to famine conditions in utero exhibited higher rates of adult-onset diseases such as obesity, diabetes, and cardiovascular disorders, compared to those unexposed.
   A landmark study by Heijmans et al. (2008) demonstrated that individuals prenatally exposed to the Dutch Famine showed persistent differences in DNA methylation at the IGF2 gene well into adulthood. These findings suggest that early-life environmental factors can induce lasting epigenetic modifications in humans. However, while such marks may persist throughout an individual's lifetime, clear evidence that these epigenetic changes are reliably transmitted to subsequent, unexposed generations in humans remains limited. The possibility of true transgenerational epigenetic inheritance in humans—such as has been documented in some animal models—remains an open and actively debated question.
   The Dutch Famine studies are thus best understood as examples of intergenerational epigenetic effects, where environmental factors during early development can have long-lasting consequences on gene regulation and health. Although intriguing, durable transgenerational inheritance in humans remains unproven; current evidence supports intergenerational marks in exposed cohorts. These studies nevertheless demonstrate how profoundly the environment can shape gene expression and, potentially, influence the health and evolutionary trajectory of human populations.
2. Epigenetics in Darwin's Finches
   Darwin's finches, often cited as a textbook example of natural selection, provide another intriguing case of epigenetics in evolution. These finches, found in the Galápagos Islands, have evolved different beak shapes to exploit different food sources. Genetic mutations are involved in this diversification; recent research has revealed that epigenetic factors also play a part in their adaptation.
   A collaborative study by researchers at the University of Utah and the Smithsonian Tropical Research Institute revealed that Darwin's finches with differing beak shapes also displayed distinct patterns of DNA

methylation. These epigenetic differences were linked to variations in both diet and habitat, suggesting that environmental conditions can influence gene expression in real-time and may play a role in the finches' adaptive potential. Notably, Skinner et al. (2014, BMC Genomics) reported epigenetic variation associated with beak differences; these findings are hypothesis-generating and do not, by themselves, establish causality.

3. Epigenetic Inheritance in Plants
   Plants provide some of the clearest examples of epigenetic inheritance. They face environmental challenges such as drought, temperature fluctuations, and soil conditions that require rapid adaptation. The model plant *Arabidopsis thaliana* has been shown to undergo epigenetic changes in response to drought conditions, for instance. A primary paper demonstrating stress-induced epigenetic inheritance in plants is by Boyko et al. (2010, *Proceedings of the National Academy of Sciences*), showing transgenerational effects of stress in *Arabidopsis*. These changes, which involve alterations in DNA methylation and histone modification, help the plant conserve water and survive in dry environments.

   In plants, stress-induced epigenetic states can sometimes persist across generations in specific lineages and experimental contexts, though persistence often attenuates over time. For example, offspring of drought-stressed plants may exhibit enhanced drought resistance even when grown under normal conditions—a phenomenon documented in experimental work (Pecinka and Mittelsten Scheid 2012). While the underlying mechanisms are still being elucidated, these findings suggest that plants can, in some cases, inherit adaptive epigenetic modifications triggered by ancestral environmental conditions. This demonstrates that epigenetic changes induced by environmental stress can be inherited, allowing plants to rapidly adapt to changing environments without waiting for random genetic mutations.

Epigenetics offers some of the strongest evidence that our genome participates in an active exchange with the environment. The DNA sequence is relatively stable, but gene expression is negotiable; it can shift with experience, resetting the organism's internal settings on physiological timescales. Mutation and natural selection still govern the long arc of evolutionary change in populations, but epigenetic regulation operates within a single lifespan, helping organisms meet immediate demands while genetic adaptation remains slower. In this view, the environment is not mere backdrop—it is a stream of information continuously sampled, filtered, and translated into biological action. When signals persist, regulatory systems can move whole programmes at once, redirecting energy and attention among growth, repair, defence, and reproduction. These are not random flickers in the genome's output but coordinated reallocations—suites of genes rising and falling together, like a rehearsed response rather than a scattered reaction. Epigenetics thus reframes adaptation as a two-level story: slow genetic change across generations, and faster regulatory control within organisms that can buffer life in the meantime, buying time for the deeper work of evolution.

2.4 The Evolution of the Eye: Information as a Guiding Force

A camera's functioning is determined by several fundamental physical formulas that govern how light is focused, how exposure is managed, and how images are formed. To understand how the eye works as a biological camera, let's briefly review the physical formulas that govern all optical systems. (The following is a simplified summary; the essential point is that both biological and artificial lenses follow these universal rules.) At the heart of a camera's optical system is the thin lens formula,

$$\frac{1}{f} = \frac{1}{d_o} + \frac{1}{d_i}$$

which relates the focal length (f) to the distances from the lens to the object (d_o) and from the lens to the image (d_i). This formula determines how far from the lens the image will be focused, a crucial element for achieving sharpness in photographs.

Building on this, the lensmaker's formula

$$\frac{1}{f} = (n-1)\left(\frac{1}{R_1} - \frac{1}{R_2}\right)$$

describes how focal length is affected by the lens's material and curvature. This formula is essential in lens design, as it enables precise control over focal length, zoom, and field of view in different types of lenses. f is the focal length of the lens, n is the refractive index of the lens material, and R_1 and R_2 are the radii of curvature of the lens's two surfaces.

The amount of light a camera can gather is controlled by the aperture, defined by the *f*-number (*f*-stop) formula,

$$N = \frac{f}{D}$$

where f is the focal length and D is the diameter of the aperture. The *f*-number determines the amount of light entering the camera and affects both exposure and depth of field. Lower *f*-numbers (e.g., *f*/1.8) allow more light and produce a shallow depth of field, ideal for portraits, while higher *f*-numbers (e.g., *f*/16) let in less light and increase depth of field, suitable for landscape photography.

The exposure triangle, which comprises aperture, shutter speed, and ISO, controls the camera's exposure. Exposure values (EVs) are calculated using the formula:

$$\mathrm{EV}_{100} = \log_2\left(\frac{N^2}{t}\right)$$

where N represents the *f*-number and t is the exposure time (shutter speed). Adjusting any part of the triangle requires adjustments in the other two to achieve a balanced exposure, ensuring that the photograph is neither too bright nor too dark.

The magnification of the camera, determined by

$$M = -\frac{d_\mathrm{i}}{d_\mathrm{o}},$$

controls how large objects appear in the image. This formula helps define the field of view, impacting how much of a scene can be captured within the frame. Additionally, the diffraction limit, described by the formula

$$\theta \approx 1.22\frac{\lambda}{D},$$

sets the maximum resolving power of a lens based on the wavelength of light (λ) and the aperture diameter (D). This limit affects the sharpness of images, especially at small apertures.

Finally, the inverse square law

$$\mathrm{I} \propto \frac{1}{r^2}$$

explains how light intensity falls off with distance from a source. This principle is vital for understanding lighting in photography, as light weakens over distance, affecting how subjects are illuminated. Together, these formulas describe the essential physics behind camera function, shaping how light is gathered, focused, and managed to create clear, well-exposed images.

The foundational laws of lenses—core principles that reveal the mysteries of light and vision—are the bedrock of optical science, studied and debated in classrooms and research labs around the world. These principles form the backbone of advanced physics and engineering, shaping our understanding of everything from microscopes to telescopes. But a deeper, more astonishing question lies beyond the textbooks: How did nature, with no blueprint or design, shape the eye—a biological camera that follows these very laws with breathtaking precision? The precise coordination required for structures like the lens and cornea to function together and focus light raises questions about whether randomness alone could account for such intricate design. How did the genetic blueprint "know" the laws of optics to create such an efficient visual system? Of course, there was no foresight or intention—only the slow accumulation of changes filtered by the unyielding demands of physics and function. Yet the emergence of the eye remains one of nature's most remarkable convergences between biology and physical law.

The evolution of the eye reveals not a weakness of Darwinian theory, but the necessity to see it in a broader, modern light. While random mutations do introduce new genetic variation, the assembly of an organ as complex and functionally exquisite as the camera eye cannot be explained by chance alone. Instead, the story is one of cumulative selection, structural constraints, and deep informational feedback across generations—each tiny change tested against the requirements of light, focus, and image formation. As classic research by Nilsson and Pelger (1994) demonstrated, the gradual evolution of the vertebrate eye can be plausibly modelled as a series of incremental, selectable steps—each conferring a marginal improvement in vision and thus favoured by natural selection. This process is not random in its outcome, but powerfully shaped by the physical constraints of optics, as also discussed by Land and Nilsson (2002).

Moreover, the independent evolution of complex eyes in both vertebrates and cephalopods is a textbook case of convergent evolution, demonstrating that certain physical solutions are "discovered" repeatedly by life under similar constraints—as Ogura et al. (2004) show in a genomic comparison of these lineages. What appears improbable when viewed as a single leap becomes inevitable when seen as a long chain of viable, information-rich adaptations, each guided by feedback from the environment.

In this sense, the "knowledge" embedded in the genetic blueprint is not conscious understanding, but the accumulated information of survival—an archive of successful solutions to the universal problems of light and vision, written not in words, but in the structure and function of biological tissue. "The eye, in all its intricacy, is not the product of randomness, but of natural selection operating over vast timescales, with each refinement ratified by the requirements of physical law" (Nilsson 2013). The eye functions as a sophisticated optical system. The cornea and lens focus light onto the retina, where photoreceptors convert it into electrical signals for the brain. This process requires precise refraction, accommodation, and biophysical mechanisms. At the molecular level, the crystalline proteins in the vertebrate lens are arranged with astonishing precision to maximise transparency and minimise light scattering—an adaptation whose effectiveness can be explained by the relentless sieve of selection, as demonstrated in studies of protein evolution in the eye (Wistow and Piatigorsky 1988). These proteins, derived from gene duplication and co-option of metabolic enzymes, illustrate how evolution can recruit existing molecules to serve new optical functions, refining them incrementally to suit the physical demands of vision. This is not evidence of foresight, but of what Francois Jacob called "evolution as a tinkerer, not an engineer" (Jacob 1977), assembling complex systems through the opportunistic reuse and gradual improvement of available parts.

This remarkable alignment extends into the developmental choreography of the eye itself. During embryogenesis, molecular gradients and signalling pathways—such as the *Pax6* gene-regulatory network—guide the precise formation of cornea, lens, and retina, often independently in separate lineages. The celebrated *Pax6* gene, dubbed the "master control" gene for eye development, is found across animals as distant as fruit flies and mammals, a testament to deep homology and to the convergent recruitment of genetic tools in response to similar selective pressures (Gehring 1996; Carroll 2008). *Pax6* gene shows deep homology across animals, though it is one node in a broader regulatory network rather than a sole determinant. These processes do not consciously "know" the laws of optics; rather, their development is shaped by the iterative interplay between random genetic variation and the consistent selective force of light itself. The presence of light in the environment acts as both a stimulus and a filter, continuously favouring even the simplest mutations that improve light detection. Over countless generations, this persistent exposure to light channels evolutionary possibilities, gradually sculpting photoreceptive structures that align ever more closely with optical principles. While the genetic code encodes the molecular instructions for protein synthesis, it cannot alone account for the exquisite coordination and adaptability seen in eye formation. Instead, information flows at every level—molecular, cellular, and systemic—ensuring that biological systems evolve and develop in harmony with the physical realities of their environment. In this sense, the very fact that light stimulates the evolution of the eye points to a deeper informational dynamic at work in evolution itself.

Nature's selection of optical materials in the eye—such as the aqueous and vitreous humours—offers further evidence of how physical constraints shape evolution. These fluids, with specific refractive indices, help optimise light focus; the ordered packing of lens crystallins minimises scattering and preserves transparency. This is a level of precision that arises from cumulative selection acting on variation under optical constraints; no foresight is required for incremental improvements to accumulate. In this case, physical law sets the problem—how to transmit and focus light with minimal loss—while selection acts as an information-accumulating process. Over generations, variants that better satisfy the optics (e.g., appropriate refractive indices, low-scatter protein packing) are retained, so the eye's "design" reflects stored information about those constraints rather than foresight.

Modern evolutionary theory recognises that while mutation supplies raw variation, the direction and intricacy of adaptations are strongly guided by the interplay of genetic constraints, physical law, and the feedback-rich dynamics of development.

Modern evolutionary theory recognises that while genetic mutation provides the raw material for variation, the direction and complexity of adaptations are powerfully shaped by the interplay of genetic constraints, physical law, and the feedback-rich structure of developmental systems. The role of "informational principles" in evolution is explored in both theoretical and empirical work: Maynard Smith (2000) provided an early framework for thinking about biological information, and Adami (2016) reviews how information theory quantifies evolutionary adaptation. However, it is not information per se that guides evolution, but the fact that selection preserves those configurations—genetic, developmental, and morphological—that best match the constraints and opportunities presented by the physical environment.

Evolution is not a purely random process: random variation is shaped by non-random selection and constraints from physics, development, and ecology. Genetic instructions and environmental feedback, when combined, can serve as a guiding force that shapes the course of evolution. This aligns with ideas from systems biology, which views living organisms as highly organised systems that process information, adapt, and self-organise in response to their environment. As information becomes critical here, it is not only genetic code but also a universal principle that organises energy and matter according to specific laws. The evolution of the eye may thus be viewed as an information-driven process where the physical principles required for vision are encoded into evolution.

Ultimately, the human eye, with its intricate design and alignment with the laws of physics, illustrates nature's ability to organise matter into highly efficient systems. The process raises profound questions regarding the ability of biological evolution to create such precise, fine-tuned structures. When considering information in this context, it arises not as a passive descriptor but as an active force that guides both biological evolution and the fundamental structure of the universe.

2.5 Analogous Organs: Convergence in Evolution and the Role of Information

Imagine nature as a series of independent scientists, each experimenting, trial after trial, to create a design that defies gravity and allows life to take to the skies. In the case of birds, bats, and insects, we see three evolutionary "inventions" of flight, each arising separately, yet arriving at similar principles of lift and thrust. The astonishing convergence of winged forms across these lineages mirrors the discoveries of human engineers in flight, who, studying lift and aerodynamics, came to understand the fundamental mechanics necessary for any machine—or animal—to soar.

For birds, flight emerged through the transformation of forelimbs into feathered wings. Over millions of years, evolution sculpted bones, muscles, and feathers into an efficient, lightweight structure designed to generate lift. Bat wings evolved differently; they adapted a delicate membrane stretched between elongated fingers, creating a flexible, highly manoeuvrable wing. Insects, meanwhile, evolved an entirely different structure, with rigid wings attached to powerful muscles in the thorax, allowing them to generate the rapid, repeated beats necessary to maintain lift in their tiny bodies.

So how did nature "discover" the physics of flight? Like human engineers who tested wing shapes and airflow in wind tunnels, natural selection acted as a relentless testing ground. Over countless generations, mutations that resulted in slight increases in lift, control, or efficiency were rewarded with better survival and reproductive success. Each lineage "experimented" within its own anatomical constraints, and those that achieved stable flight were favoured, honing winged designs that mirror the mathematical elegance of human aerodynamics.

Comparing natural flight to our understanding of aerodynamics, we find remarkable alignment. To achieve lift, wings need a surface that curves, creating a pressure differential that generates upward force. Birds' wings, with their asymmetrical cross-section and smooth, cambered feathers, operate on this principle, channelling airflow to achieve stable lift and allow for gliding and soaring. Bat wings, though membrane-bound, also have a cambered shape, adjusting to provide lift with flexible articulation. Insects, too, follow the basics of lift through rapid wing beats that create vortices above their wings, achieving a form of lift remarkably similar to helicopter rotors.

The marvel lies in the idea that, without an understanding of physics, evolution stumbled upon these principles through the relentless push of natural selection. Nature didn't have a blueprint for the laws of aerodynamics but arrived at them nonetheless through trial, error, and adaptation. Each wing type converges on the same core principles—lift, thrust, and control. What took scientists centuries to understand was refined by nature's experiments over millions of years, creating creatures as diverse as hummingbirds, moths, and bats that each embody the mathematical elegance of flight.

To reflect on this is to realise that nature itself is a master engineer. The repeated discovery of flight through convergent evolution highlights how the principles of physics are universal, embedded in the very fabric of the world. Evolution is not only a process of adaptation but a form of refinement of alignment of information between the living organism and its surroundings, where organisms explore and exploit the laws of nature to survive. By finding these same principles, nature appears to demonstrate a form of "adaptive intelligence"—not conscious planning, but a process of selection and constraint that leads to solutions echoing those that humans later discover through science. The wings of birds, bats, and insects are a testament to the universe's inherent patterns, with life finding solutions that resonate with the principles humans would later uncover—an awe-inspiring example of the universality of information.

Analogous organs, or analogous structures, are defined as organs that perform similar functions but have different evolutionary origins. This contrasts with homologous organs, which share a common evolutionary ancestor but may serve different functions in modern species. Analogous organs arise through a process known as convergent evolution, where unrelated species evolve similar traits independently due to similar environmental pressures and functional needs.

The existence of analogous organs—structures with similar functions but independent evolutionary origins—demonstrates how physical laws and environmental constraints channel the process of evolution, leading to repeated "solutions" across distant lineages. The phenomenon of convergence is superbly documented in comparative genomics (Parker et al. 2013; Ogura et al. 2004) and elegantly synthesised in Conway Morris's *Life's Solution* (2003).

This is not a claim that evolution is deterministic or preordained but that the space of viable solutions is bounded by the "rules" of physics, chemistry, and developmental possibility (Kirschner and Gerhart 1998). These laws set boundaries on what is possible in the biological world, often leading to similar solutions in organisms that face similar challenges.

One of the primary driving forces behind the evolution of analogous organs is the similarity of environmental pressures faced by different species. For instance, animals that live in aquatic environments tend to develop streamlined bodies and fins or flippers to move efficiently through water. Birds and insects that rely on flight tend to develop wings that share common aerodynamic properties, despite significant differences in their evolutionary history.

These environmental pressures create selective forces that favour certain adaptations over others. While the specific mutations that occur in different species are random, the environmental context in which these mutations are selected is not. As a result, species living in similar environments often evolve similar solutions to survival problems, even if they are not closely related.

Physical laws, such as those governing fluid dynamics, gravity, and energy efficiency, also play a significant role in shaping the evolution of analogous organs. For instance, the principles of aerodynamics dictate that any organism that relies on flight—whether it is a bird, a bat, or an insect—must have wings that create lift while minimising drag. The specific structures of bird wings and insect wings differ greatly, but they must both adhere to these aerodynamic principles to function effectively.

Similarly, by affecting the evolution of aquatic animals, the physics of buoyancy and drag play a role. Whether examining fish, dolphins, or penguins, we see the development of streamlined bodies that reduce resistance as these animals move through water. This is not because they share a common ancestor with streamlined bodies, but because the physical laws governing movement in water impose similar constraints on their body shapes.

In this way, physical laws act as a form of "information" that guides the evolutionary process. They set boundaries on what is possible and shape the direction that evolution takes, leading to the development of analogous organs in species that face similar environmental challenges.

Case Studies of Analogous Organs

a. Wings of Birds, Bats, and Insects

One of the most famous examples of analogous organs is the evolution of wings in birds, bats, and insects. All of these species evolved the ability

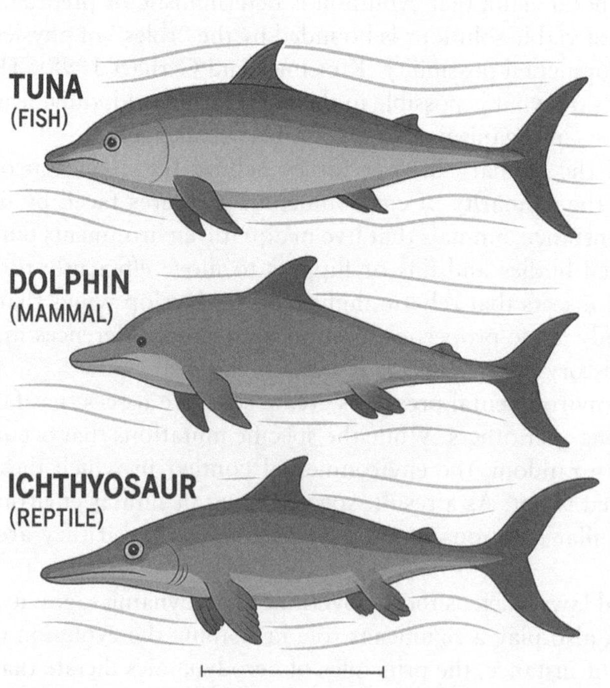

FIGURE 2.2 Streamlining in water: tuna (fish), dolphin (mammal), and ichthyosaur (reptile) showing convergent fusiform shapes that reduce drag; note differing tail designs.

to fly, but their wings are structurally very different. Bird wings are modified forelimbs covered with feathers, while bat wings are made of skin stretched over elongated finger bones. Insects, on the other hand, have wings that are extensions of their exoskeleton.

Despite these structural differences, the wings of birds, bats, and insects all serve the same function: flight. The similarities in shape and function of these wings result from the physical laws governing aerodynamics. Any organism that evolves the ability to fly must develop wings that create lift and minimise drag, leading to the convergent evolution of similar wing shapes in these different species.

b. Fins of Fish and Flippers of Marine Mammals

Another well-known example of analogous organs is the evolution of fins in fish and flippers in marine mammals such as dolphins and whales. Fish and marine mammals are not closely related in evolutionary terms, yet both evolved similar structures to move efficiently through water. Fish have fins used for propulsion and steering, while marine mammals have flippers that serve the same purpose.

The evolution of these analogous organs is driven by the physical properties of water. To move efficiently in an aquatic environment, animals must reduce drag and increase propulsion. This leads to the development of streamlined bodies and appendages that generate thrust, while reducing resistance, regardless of their evolutionary history.

c. Eyes of Vertebrates and Cephalopods

The eyes of vertebrates (such as humans and birds) and cephalopods (such as squids and octopuses) provide another striking example of analogous organs: both possess a lens that focuses light onto a retina with photoreceptor cells, yet these complex camera eyes evolved independently. Research by Ogura, Ikeo, and Gojobori (2004, *Proceedings of the National Academy of Sciences*) compared vertebrate and cephalopod eye development at the genetic level, highlighting the independent evolutionary origins of these remarkably similar structures. Vertebrate eyes develop from an outgrowth of the brain, while cephalopod eyes develop from a different part of the embryo.

Despite their different evolutionary origins, the camera-type eyes of vertebrates and cephalopods display striking similarities in structure and function—each possessing a lens that focuses light onto a retina lined with photoreceptor cells. This remarkable convergence is not the result of shared ancestry but the repeated, independent arrival at solutions imposed by the physical constraints of image formation in watery worlds. Recent genomic and developmental studies demonstrate that while these eyes utilise analogous genes and developmental pathways, their similarities arose independently through convergent evolution, rather than from a common ancestral eye (Ogura, Ikeo, and Gojobori 2004). This highlights the principle that, under similar selective pressures, evolution can repeatedly discover functionally optimal structures—even when starting from entirely different genetic blueprints.

d) Echolocation in Bats and Dolphins
Echolocation provides another fascinating example of convergent evolution, where different species developed analogous organs for similar functions. Both bats and dolphins evolved the ability to use echolocation to navigate and hunt in environments where vision is limited. Bats use high-frequency sounds to detect objects in the dark, while dolphins use similar sound waves to detect objects in murky water. Both bats and dolphins independently evolved the ability to use echolocation for navigation and hunting in environments with limited visibility. Parker et al. (2013) conducted a pivotal genomic analysis revealing molecular convergence in hearing-related genes between these two groups, providing compelling evidence for the genetic underpinnings of their similar echolocating abilities.

The structures involved in echolocation are quite different in bats and dolphins. Bats have specialised laryngeal muscles that produce ultrasonic sounds, while dolphins use their nasal passages to create clicks. Despite these differences, the result is the same: both species can "see" their environment through sound, thanks to the principles of acoustics.

In addition to environmental pressures and physical laws, developmental constraints play an important role in shaping the evolution of analogous organs. These constraints refer to the limitations imposed by an organism's genetic and developmental pathways, which influence the direction that evolution takes.

For example, vertebrates are constrained by the fact that their limbs develop from a specific set of genes and tissues. As a result, vertebrate wings—whether belonging to birds, bats, or pterosaurs—are always modified versions of the same basic limb structure. Insects, however, have a completely different developmental pathway for their wings, leading to a very different wing structure.

Developmental constraints do not prevent the evolution of analogous organs, but they do influence the specific forms these organs take. In some cases, these constraints may limit the range of possible adaptations, while in others, they may lead to novel solutions not seen in other species.

The phenomenon of convergent evolution reveals something astonishing: that evolution is not merely a blind wander through random mutation, but a deeply patterned process—channelled, constrained, and sculpted by the informational architecture of reality itself. This is well-documented in the literature on convergent evolution, where similar environmental pressures lead to analogous adaptations across unrelated lineages. The laws of physics, the principles of chemistry, and the conditions of survival do not simply set the stage; they impose constraints and offer solutions. Across distant lineages and unrelated species, life independently arrives at similar answers—not by chance, but because those answers are etched into the very structure of the world.

Nowhere is this more evident than in the case of aquatic mammals. Whales, dolphins, seals, manatees, and even semi-aquatic mammals like walruses evolved from terrestrial ancestors that once walked on land. As the ancestors of these animals increased in size and bulk, the biomechanics of terrestrial life began to present profound structural challenges. On land, gravity is an unrelenting force—every kilogram of body mass exerts stress on the joints, tendons, ligaments, and bones that support it. For large mammals, the simple act of standing becomes a feat of constant muscular and skeletal compensation. The limbs, particularly the distal joints and foot structures, must endure repeated high-impact forces during locomotion. Cartilage must absorb shock with minimal room for error, and ligaments are strained to maintain both flexibility and stability. As body size increases, these biomechanical stresses do not scale linearly—they increase exponentially due to the square-cube law: while volume (and thus mass) grows with the cube of an animal's linear dimensions, surface area—including the cross-sectional area of load-bearing bones—grows only with the square. This means that larger animals are disproportionately more burdened by their own mass.

This mismatch places immense strain on structural integrity and metabolic sustainability. The energy expenditure required to move, stand, and thermoregulate scales up with body mass, but the structural support does not keep pace. As a result, evolution faced a bottleneck: How could increasingly large terrestrial animals continue to function efficiently without collapsing under their own weight or expending unsustainable amounts of energy just to remain mobile?

Water provided a radical solution.

Aquatic environments offered a medium where buoyancy—Archimedes' principle in action—could dramatically offset the effects of gravity. The upward force exerted by displaced water reduced the mechanical load on limbs and joints, effectively providing structural support "for free." In this way, the water acted as a compensatory scaffold, allowing larger animals to maintain volume and mass without incurring the escalating energy and mechanical costs associated with terrestrial life. This transition allowed for a reallocation of biological resources: limbs could shrink or repurpose, axial musculature could adapt to new modes of propulsion, and metabolic energy could shift from posture maintenance to more efficient movement through the water.

Indeed, many of today's aquatic mammals—whales, manatees, seals—trace their ancestry to land-dwelling creatures that once bore the full burden of gravity. Their return to water wasn't merely an ecological detour but a strategic move towards energy efficiency, an evolutionarily intelligent recalibration of energy and form. In this view, the move to water was not accidental but a form of natural problem-solving, guided by physical laws and energy constraints. It is evolution behaving like an engineer—experimenting, assessing, and selecting solutions based on the environmental calculus of efficiency. The full fossil sequence of whale evolution from terrestrial artiodactyls to fully aquatic whales was established by Thewissen et al. (2001).

This shift to water didn't happen all at once. Early aquatic mammals, like the semi-aquatic ancestors of cetaceans, likely waded in shallow rivers or coastal zones. But over time, environmental constraints of high-density medium enforced genetic changes on these animals to align with water's hydrodynamic properties—more paddle-like limbs, elongated bodies, denser bones (to avoid excessive floating), and eventually the development of flukes for axial propulsion. In essence, these animals discovered buoyancy as a strategy, the way humans later discovered fire: not by conscious analysis, but by interaction and feedback with their environment, reinforced by the survival benefits conferred.

From a bioenergetic perspective, the transition to aquatic life provided profound metabolic advantages. Comparative studies in animal physiology consistently show that aquatic locomotion, particularly in large-bodied organisms, is significantly more energy-efficient than terrestrial movement. Locomotion on land involves repeated lifting and repositioning of limbs against gravitational force, resulting in substantial mechanical work with each stride. Muscles must not only generate forward thrust but also counteract vertical loading forces, leading to increased metabolic demand.

In contrast, aquatic propulsion—especially in species with streamlined morphologies and high-aspect-ratio caudal flukes—enables movement via undulatory or oscillatory motions that harness continuous hydrodynamic interaction with the surrounding fluid. Instead of lifting the body, aquatic animals displace water by transmitting wavelike muscular contractions along their axial musculature, producing forward thrust through the conservation of momentum. This form of propulsion minimises energy loss, especially in laminar flow regimes, and leverages the density of water to generate efficient lift and thrust with minimal vertical displacement. Fish (1996) explains the transition in mammalian swimming from drag-based propulsion to lift-based oscillatory propulsion, a key biomechanical shift underlying streamlined aquatic locomotion.

Quantitative studies in marine mammalogy have demonstrated that cetaceans, such as whales and dolphins, can traverse vast oceanic distances with remarkably low energetic cost. In fact, on a per-kilogram-per-kilometre basis, the energy required for a whale to swim is often less than the energy many large terrestrial mammals expend simply maintaining posture or standing still at similar scales; buoyancy reduces support costs, and hydrodynamic streamlining lowers transport costs. This extraordinary efficiency underscores why aquatic re-adaptation was not only viable but evolutionarily advantageous, converting a biomechanical burden into a fluidic solution optimised by the physics of buoyancy and flow.

Buoyancy also helps with thermoregulation. Large bodies submerged in water lose heat more rapidly, but aquatic mammals evolved thick blubber layers—another adaptation enabled by energy savings from buoyancy. This thermal insulation helps them maintain internal temperature with lower energetic input than if they had to constantly adjust via muscular shivering or metabolic compensation on land.

Thus, what began as an ecological necessity evolved into a complete informational alignment with the physics of water. Buoyancy acted as a silent ally, favouring morphologies and metabolic efficiencies that terrestrial life could never afford. It wasn't mere randomness that drove these species to reshape themselves in the image of fish; it was the relentless refinement of form to meet the constraints of fluid dynamics. Gravity had once sculpted their limbs for land. Now, water was re-engineering their bodies for buoyant efficiency. Beneath this transformation lies a deeper principle: evolution as a system not only responding to selective pressure but actively aligning itself with the informational structure of its environment. From this perspective, natural selection is not the sole sculptor—information itself becomes a guiding force, illuminating a dynamic dialogue between organism and environment, between constraint and innovation. Here, "guiding force" should be read as a metaphor for emergent constraint: evolutionary processes are shaped by the regularities of physics, chemistry, and information, not by any intentional agency.

Physical laws, developmental constraints, and environmental pressures are not merely external forces; they are streams of information that continuously shape and refine the living world. This information—whether it's the principles of aerodynamics, resource availability, or habitat-specific pressures—serves as a blueprint, guiding evolutionary processes towards forms and functions optimised for survival.

In this framework, "information" includes the regularities and constraints that shape an organism's development. Genetic information, passed down

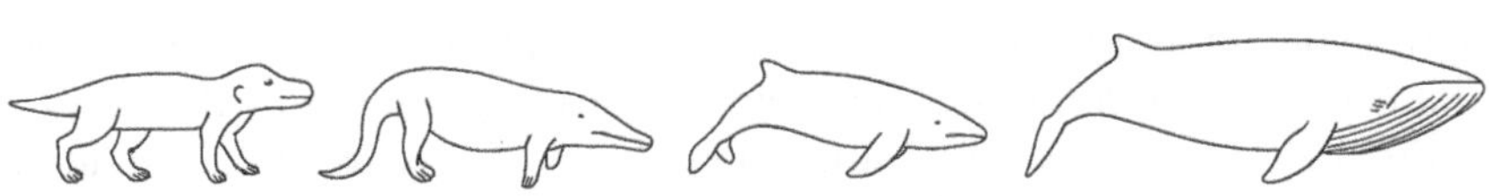

FIGURE 2.3 Whale evolution sequence (left to right): Pakicetus, Ambulocetus, Dorudon, Modern whale. Black-and-white line diagram.

through generations, acts as an adaptive record that responds to the physical world, gradually refining itself to better align with these external demands. Natural selection acts as the mechanism for this refinement, constantly "testing" genetic configurations against the world's constraints. Successful traits are preserved, while less favourable ones are discarded, leading to a genetic architecture that increasingly mirrors the enduring principles of physics and environment.

2.6 Framing Evolution: Engaging with the Theories that Shape the Modern View

To advance this view—that evolution is a recursive, information-optimising process—we must examine how it resonates with, diverges from, and enriches the major schools of evolutionary theory that have shaped our understanding of life. What is being proposed here both resonates with and challenges established schools at the intersection of biology and information science. In what follows, we shall explore how this informational perspective aligns with—and sometimes departs from—advances in evolutionary developmental biology, biosemiotics, and complex adaptive systems theory, as well as how it enters into dialogue with structuralist and emergentist philosophies of life.

Any attempt to reimagine evolution through the lens of information must engage with the powerful intellectual currents that have shaped the life sciences over the past century. This chapter's thesis—that evolution is a recursive, information-optimising process—is indebted to, and stands in conversation with, a lineage of theories that illuminate how life organises itself, adapts, and discovers.

Evolutionary developmental biology, or "evo-devo," has been a primary engine of this transformation. Building on the pioneering insights of John Maynard Smith and Mary Jane West-Eberhard, evo-devo has revealed that the "search" for evolutionary novelty is not random but constrained and enabled by developmental pathways and plasticity. In West-Eberhard's seminal book *Developmental Plasticity and Evolution* (2003), she argues that environmental signals and organismal responses are not mere accidents, but fundamental drivers of evolutionary innovation—a thesis foreshadowed by Maynard Smith et al.'s (1985). Evo-devo thus recasts the genome not as a static code but as a malleable system, open to environmental "information" and capable of accommodating, even anticipating, ecological change. The notion of evolution as information flow draws deeply from this tradition, seeing adaptation not as blind chance but as the canalisation of informational gradients.

Equally compelling is the field of biosemiotics, which explores the processes of meaning-making and sign interpretation in living systems. Scholars such as Jesper Hoffmeyer and Marcello Barbieri have advanced the influential idea that life, at every level, operates as a system of signs and codes—"messages" that are interpreted and acted upon by cells, organisms, and even ecosystems. As discussed by Barbieri (2015) and by Hoffmeyer (2008), evolution is shaped not merely by shifts in gene frequencies but by the emergence of complex semiotic networks. These networks allow living systems to "read" and "write" the informational scripts of survival, continually negotiating meaning in response to environmental cues.

Framing evolution as a feedback-driven information system is greatly enriched by this semiotic perspective. Adaptation is thus understood not only as a physical adjustment but also as an interpretive process—a continuous negotiation of meaning between organism and environment.

The metaphor of nature as a computational engine owes much to the field of complex adaptive systems. Stuart Kauffman's *The Origins of Order* (1993) and Simon Levin's classic essay "Ecosystems and the Biosphere as Complex Adaptive Systems" (1998, Ecosystems) both articulate a vision of life as a dance of self-organisation and emergence, where simple rules at the micro-level give rise to astonishing diversity and order at the macro-level. Kauffman's idea of the "adjacent possible" describes evolution not as a path blindly groping in the dark but as a creative algorithm constantly sampling new configurations, expanding the space of the possible through informational feedback. Levin demonstrates how feedback between individuals and the environment yields complex, stable patterns at higher scales—resonating with this chapter's view of evolution as the recursive optimisation of information flow.

At the interface of science and philosophy, Terrence Deacon and Denis Noble have called for a more nuanced epistemology of information in biology. Deacon (2011) contends that living systems are shaped not only by their physical components but by emergent properties—constraints, absences, and possibilities arising from the complex organisation of matter and energy. Similarly, Noble (2006) challenges the reductionist view of the "selfish gene," highlighting instead the dynamic interplay of multiple levels—genes, cells, organs, and organisms—each engaged in an ongoing exchange of information. From these philosophies emerges the concept of an "informational fabric": not a case for external design, but a recognition that the layered, self-organising principles of biological systems give rise to emergent structures with genuine causal power. Organisation itself becomes a force, both a product of evolutionary history and a determinant of future possibility, shaping the trajectory of life through the recursive play of structure, constraint, and information.

This orientation inevitably raises a profound question: Is the "informational fabric" merely a metaphor, or does it point to a real ontological structure underpinning biological order? In addressing this, the argument finds resonance with both structuralist and emergentist traditions. Structuralists—such as Günter Wagner, whose work on the biological homology concept (*Annual Review of Ecology and Systematics*, 1989) explores how inherent regularities, from physical laws to developmental pathways and mathematical symmetries, channel and constrain the evolution of form—propose that nature's possibilities are shaped by deep structural logics. Emergentists, drawing on the insights of complexity theory, maintain that new levels of organisation—cells, tissues, minds—arise from the dynamic interplay of simpler components, producing properties that cannot be reduced to their constituent parts (Kauffman 1993; Deacon 2011). Within this framework, evolutionary "guidance" is not a gesture towards mystical teleology but an acknowledgement of the inevitability that, under given constraints, nature converges upon stable, efficient, information-rich patterns—a kind of grammar inscribed in the logic of the possible.

In sum, the perspective advanced in this chapter—evolution as information optimisation—serves as both a synthesis and a provocation. Drawing deeply from evolutionary developmental biology, biosemiotics, systems theory, and the philosophy of biology, this view brings together disparate threads into a coherent tapestry. Yet it also poses a daring question: Could it be that life is not only shaped by information, but that information itself operates as a hidden order—a principle or "algorithm" intrinsic to nature—guiding the ceaseless emergence of complexity, form, and meaning?

With this foundation established, the narrative now turns to examine how an informational lens can illuminate enduring mysteries at the heart of biology: the origins of form, the convergences that give rise to analogous structures, and the creative, adaptive search that propels life's endless experimentation. From this vantage, evolution emerges not as a solitary monologue of genes but as a continuous dialogue—a negotiation between internal genetic architectures and the external informational landscape shaped by physical law and environmental constraint. In this dialogue, physical laws and ecological pressures are not mere obstacles but persistent guides, quietly steering the evolutionary journey of life.

Over generations, genetic material adjusts to harmonise with this guiding information, gradually evolving structures, behaviours, and forms that reflect the optimal arrangement for survival within given constraints.

Thus, evolution is not a chaotic process but a sophisticated exchange between organisms and the information encoded in their surroundings. Physical laws, rather than limiting life, provide a stable foundation that life builds upon, while environmental pressures ensure that adaptation remains in tune with changing conditions. Together, these forces act as an intricate guide, steering evolution towards forms that are finely tuned to the demands and possibilities of the universe. In this way, information is more than a passive factor; it is a continuous, active force that shapes life to adapt, endure, and flourish within the boundaries set by nature's laws.

2.7 Echoes across Lineages: Informational Constraints in Analogous Evolution

Analogous organs provide a clear example of how information guides evolution. The fact that different species evolve similar structures to solve similar problems suggests that evolution is not entirely random. Information present in the environment and the genetic material of the organisms themselves shape it, instead. Analogous organs often follow predictable patterns, supporting this idea. For example, the wings of birds, bats, and insects all share certain aerodynamic features, despite their different evolutionary origins. This suggests that the physical laws governing flight impose constraints on the forms wings can take, providing a kind of "information" that guides evolution.

The evolution of analogous organs offers fascinating insight into how physical laws and environmental pressures shape the development of organisms. These organs, which evolved independently in different species yet serve similar functions, demonstrate that evolution is not merely a process of random trial and error. Organisms' development is guided by a set of constraints and regularities, resulting in predictable outcomes.

From the moment life began, it has existed in dialogue with its surroundings. Not as a passive recipient of conditions, but as a system constantly sculpted by the pressures, signals, and information streaming in from the world around it. While classical evolutionary theory placed natural selection as the central arbiter acting upon randomly mutating genes, emerging insights suggest a deeper, more reciprocal relationship: the environment is not merely filtering random variation—it is participating in the shaping of biological form.

Recent research in evolutionary developmental biology (evo-devo) provides strong evidence for this claim. Studies have revealed that developmental plasticity—the capacity of a single genome to produce multiple phenotypes in response to environmental cues—plays a pivotal role in evolution (West-Eberhard, M.J., *Developmental Plasticity and Evolution*, 2003). Rather than waiting passively for beneficial mutations, organisms exhibit context-sensitive development that "reads" environmental inputs and modulates gene expression accordingly. These environmentally induced changes can, if adaptive, become genetically stabilised over generations—a process known as genetic accommodation (West-Eberhard 2005, *Proceedings of the National Academy of Sciences*).

In this model, environmental factors serve not merely as triggers but as informational guides that shape evolutionary trajectories. Take, for example, temperature-dependent sex determination in reptiles: in species such as the American alligator and many turtles, the ambient temperature during a critical window of embryonic development determines whether the offspring will be male or female (Bull, J.J., "Sex Determination in Reptiles," *Quarterly Review of Biology*, 1980; Janzen and Phillips 2006, *Annual Review of Ecology, Evolution, and Systematics*). This is not a mere ecological curiosity, but a profound demonstration of how environmental information can directly modulate gene expression to shape developmental fate. Over evolutionary timescales, such mechanisms themselves may evolve for greater sensitivity or robustness, depending on ecological stability, effectively embedding environmental logic into the genome's regulatory architecture (Schlichting and Wund 2014).

These temperature-sensitive sex determination systems exemplify a broader principle: environmental information enters development through multiple channels—thermal, chemical, microbial, and mechanical—each capable of modulating gene expression during critical windows.

2.8 Symbiotic Intelligence: Coevolution as Informational Convergence

If temperature is one channel of environmental information, symbiosis is another: a living, biochemical interface where hosts and microbes exchange metabolites and signals that feed directly into regulatory networks.

Symbiosis is now understood to be a fundamental principle of evolution, with the hologenome theory positioning host and microbiome as a single unit of selection (Zilber-Rosenberg and Rosenberg 2008). Likewise, the role of mechanotransduction—the translation of mechanical forces into gene expression—is highlighted by Engler et al. (2006), who showed that matrix elasticity can bias stem cell lineage specification.

These discoveries underscore the point that environmental information is continuously integrated at multiple biological levels, shaping evolutionary pathways in real-time. The hologenome theory of evolution posits that the host and its microbiota co-evolve as a single unit of selection. The composition of the microbiome, in turn, is shaped by environmental exposure—diet, climate, pH, salinity, and more. In coral reefs, for instance, coral polyps rely on symbiotic algae (zooxanthellae) that confer both energy and resilience. But different strains of algae perform differently under temperature stress, and recent work shows that corals can dynamically adjust their symbiont composition in response to warming seas. This microbial reassortment allows corals to adapt in real-time to environmental stress—a rapid, plastic response subsequently filtered by Darwinian selection at the holobiont level, a form of Lamarckian flexibility layered over Darwinian processes.

The same theme holds at the molecular scale. Beyond organismal partners, the local physicochemical milieu supplies information (pH, temperature, ions) that steers macromolecular behaviour and, in turn, cellular programmes.

Even at the molecular level, environmental conditions guide the internal architecture of life. Consider protein folding, a process once thought to be strictly dictated by amino acid sequences. It is now clear that the cellular environment—pH, temperature, ion concentration—directly influences folding pathways. In extreme conditions, some organisms evolve intrinsically disordered proteins that remain flexible until bound to their target, a form of biochemical plasticity that allows adaptation without gene sequence change. This suggests that even the functional behaviour of molecules, long thought to be "hard-coded," is tuned by environmental context. Alongside biochemical and microbial signals, mechanical cues provide a third, orthogonal information stream that cells actively decode.

One of the most underappreciated forces in evolutionary biology is mechanotransduction—the conversion of physical forces into biochemical signals that shape cellular behaviour. Stem cell differentiation, for instance, is influenced by the rigidity of the substrate on which cells grow. On soft matrices (on the order of ~0.1–1 kPa), mesenchymal stem cells tend to activate neurogenic programmes; on intermediate stiffness, they show myogenic tendencies; and on stiffer matrices (tens of kPa), they more often bias towards osteogenic fates. A classic demonstration is Engler et al. (2006), which showed that matrix elasticity can bias stem cell lineage specification rather than deterministically "direct" it. Mechanical cues are sensed through integrins and focal-adhesion complexes (e.g., FAK, Src), transmitted via cytoskeletal tension and RhoA/ROCK–actomyosin contractility, and integrated with nuclear mechanosensing (including YAP/TAZ localisation and chromatin accessibility), thereby translating physical context into gene-regulatory responses. These effects are context-dependent and interact with soluble factors and cell history, but they robustly illustrate how mechanics participate in fate decisions. In developing embryos, the physical compression of cells in a confined environment can alter the fate of those cells—a direct example of structure influencing instruction. Over evolutionary timescales, lineages can shift morphology and tissue architecture—not by "new genes" alone, but by evolving the strength, timing, and sensitivity of mechano-responsive pathways and extracellular matrices, thereby embedding mechanical logic into regulatory architecture (cf. developmental bias and plasticity frameworks).

In plants, the evidence is even more striking. Plants do not merely passively respond to light and gravity—they measure them with high precision and reroute growth patterns accordingly. The shade-avoidance response, for instance, alters internode elongation based on red-to-far-red light ratios, allowing plants to avoid being overgrown. Root branching is governed by nutrient gradients in the soil, and plants grown in heterogeneous soils develop asymmetric root systems that optimise resource uptake. These behaviours are not random; they are algorithmic-like, guided by external informational gradients and internal feedback loops. Evolution has thus equipped plants with sophisticated sensing systems that allow them to adapt plastically and evolve directionally in environments that reward perceptual acuity. A key study in this area is Forde and Lorenzo (2001), showing how nutrient heterogeneity drives root plasticity.

Across these scales, a single pattern recurs: environmental information is integrated into gene regulation through multiple channels. Thermal regimes bias sex determination; symbiotic partners modulate biochemical signalling; physicochemical context tunes protein behaviour; and mechanical fields shape cell fate. Evolution can then stabilise, amplify, or dampen these inputs by adjusting regulatory sensitivities and tissue architectures. In this sense, environments do not merely "select" outcomes—they also help specify them—and genomes record that logic in their control systems over time.

2.9 Epigenetic Echoes: Inheritance beyond DNA

Environmental influence has been reported, in specific animal models, to produce epigenetic states that appear in descendants without changes to the DNA sequence. In mice, exposure to a particular odour paired with shock induces aversion; some studies report heightened sensitivity to that odour in F1–F2 offspring alongside sperm methylation/small RNA differences. Overall, the evidence is model-dependent and mixed, and mechanisms and reproducibility remain debated; claims of robust mammalian transgenerational inheritance should be treated cautiously. These findings blur—but do not erase—the line between experience and inheritance, illustrating a possible route by which environmental information can reach the germline under certain conditions (e.g., Dias and Ressler 2014, with subsequent critiques and mixed replications).

If such effects truly reach descendants without DNA sequence change, the inherited substrate cannot be a new set of genes, but a regulatory state—a pattern that alters which genes are accessible, when they are activated, and how strongly they are expressed. Epigenetic inheritance, in this sense, is not inheritance of new code but inheritance of a bias in how existing code is read. During the formation of germ cells and again after fertilisation, epigenetic marks are extensively reset, as if the system insists on beginning each generation with a clean regulatory slate. This is precisely why claims of robust mammalian transgenerational inheritance remain contentious: the default expectation is erasure. Any genuine multi-generation effect must therefore involve signals that either escape resetting or are re-installed with high fidelity in the next generation.

But in plants and some invertebrates, transgenerational transmission of non-DNA inheritance is well documented and reproducible. In plants, a classic case is paramutation, in which one allele heritably alters the expression state of another allele without changing nucleotide sequence. The key point is that what persists is not a new gene, but a regulatory configuration—a stable pattern of chromatin marking and gene accessibility that can be transmitted through reproduction. The resulting expression state can behave as a heritable trait over multiple generations, demonstrating that gene regulation itself can be an inherited variable. In the nematode C. elegans, transgenerational inheritance is strongly supported through small-RNA–mediated gene silencing. Here, small RNAs can initiate sequence-specific silencing and, through amplification and chromatin-based reinforcement, maintain that silencing across multiple generations. This provides a mechanistic demonstration of inheritance beyond DNA sequence: regulatory molecules and the chromatin states they help establish can carry forward a gene-expression bias into descendants.

These examples clarify what "inheritance beyond DNA" means in concrete terms. Epigenetic inheritance functions as a low-bandwidth inheritance channel: it does not transmit new genes, but can transmit persistent regulatory biases—such as altered gene accessibility or silencing—into the next generation. In this way, limited information about prior conditions can be carried forward alongside DNA, shaping how the same genome is expressed in descendants.

2.10 Osmosis and the Architecture of Necessity

The discovery of osmosis as a principle of molecular movement predated our understanding of the kidney's use of this process, but the kidney's physiology demonstrates that life had already "discovered" and optimised this principle long before humans understood it. The process of osmosis—where water molecules pass through a semipermeable membrane from a region of lower solute concentration to a region of higher solute concentration—was first scientifically recognised in 1748 by Jean-Antoine Nollet. Nollet observed that when water and alcohol were separated by an animal bladder membrane, the water moved towards the alcohol, demonstrating an early understanding of selective permeability. This principle laid the groundwork for understanding how substances move across biological membranes. It wasn't until the 19th century that the physiology of the kidney began to be explored in depth. Carl Ludwig, a German physician and physiologist, played a pioneering role in the 1840s with his research into kidney function, positing that filtration in the kidneys was driven by blood pressure. Later studies in renal physiology continued to elaborate on this concept, particularly in the work of Ernest Starling, who introduced the principles of capillary filtration and reabsorption in 1896. In the early 20th century, more details of nephron structure and function, including the roles of glomerular filtration and tubular reabsorption, were identified, providing a comprehensive view of how the kidney uses osmosis and active transport to filter blood and regulate body fluids.

The sophistication of the kidney's osmotic processes prompts questions about the role of evolution in "discovering" such principles. How do genetic mutations, through trial and error, gradually arrive at the complex network of glomerular filtration and tubular reabsorption? This brings us back to the classical philosophical debate of determinism versus indeterminism: whether evolution is just random mutation or a guided process aligning with universal principles. The debate of determinism versus indeterminism offers a fascinating lens for exploring evolution. Determinism implies that biological structures, like the kidney, evolve towards predictable, inevitable outcomes shaped by universal laws. In contrast, indeterminism would suggest that these structures emerge through random genetic mutations, where chance alone drives the development of complex systems like osmosis-based filtration in the kidneys.

The kidney's sophisticated filtration system—relying on the natural principle of osmosis—emerged to support the regulation of water, salts, and waste products in large, multicellular organisms. Yet osmosis itself is not a biological invention. It's a fundamental physical phenomenon, occurring whenever water moves across a semipermeable membrane in response to solute gradients, and has been present on Earth since water and dissolved solutes first existed, long before life emerged.

If we view evolution from an indeterministic perspective, the kidney's complex structure would be the product of countless random genetic mutations that happened to align over time. Each mutation, independently arising and unconnected to the larger picture, would have either conferred some selective advantage or been neutral, accumulating gradually over generations. Eventually, this accumulation could have produced a structure like the glomerulus and tubular system that exploits osmosis as an ideal mechanism for blood filtration. According to this view, no foresight or knowledge of osmosis as an optimal solution was required; rather, the kidney's structure is simply a highly successful arrangement selected through trial and error.

However, this indeterministic view raises a compelling question: How could random genetic changes "stumble upon" such a precisely optimised mechanism for filtration? The kidney's structure is highly intricate, and each component—glomerulus, Bowman's capsule, and tubules—performs functions that are interdependent and finely tuned. Each mutation required to form this complex, integrated system would have needed to align in ways that supported incremental improvements, leading to an efficient filtering organ capable of sustaining a large organism. The complexity of this evolutionary path, with each small change needing to fit into the larger functional system, challenges the notion that sheer randomness could have arrived at this solution without some form of alignment with underlying physical laws.

From a deterministic viewpoint, the kidney's development might reflect an inherent tendency of evolution to align biological forms with pre-existing physical principles. If osmosis is the most efficient and feasible way to filter substances across a membrane, then any evolving system tasked with waste removal and fluid balance might inevitably align with osmotic principles. Evolution, in this sense, could be seen as an "informational discovery" process where organisms naturally converge towards structures that resonate with universal laws.

This deterministic perspective suggests that the kidney's structure isn't purely a product of chance but an alignment with the "information fabric" of the universe. Just as water naturally flows downhill, macromolecules and cells may naturally arrange themselves into structures that best exploit principles like osmosis. In this view, osmosis would be the most effective and inevitable solution for blood filtration, and evolution would be guided—not in a conscious sense, but by the selective pressures that inherently favour systems utilising universal physical laws.

If we accept the notion that the universe contains an inherent informational structure, then evolution may be more than just a random walk through possible configurations. Instead, it could be a process of discovery that "finds" optimal designs by aligning genetic variations with the physical properties of reality. In the case of kidneys, genetic mutations that led to semipermeable barriers and osmotic gradients within primitive organisms may have been more likely to persist because they aligned with a natural, pre-existing method for fluid regulation and waste removal.

The concept of an "information fabric" guiding evolution suggests that there are certain stable, efficient ways of organising matter to achieve specific functions—like filtering blood in a multicellular organism. In this view, the glomerulus, tubules, and osmotic gradients that form the kidney's filtration system might be seen as "discoveries" embedded in the structure of the universe. Just as physical laws shape the behaviour of matter on a cosmic scale, they may also influence the arrangement of molecules, cells, and organs to favour certain biological architectures. Evolution, then, could be understood as a process that brings life forms into alignment with these physical regularities.

Given that osmosis is a highly efficient, passive process for moving water and regulating solute balance, it's plausible that any organism requiring waste filtration and fluid balance would eventually evolve structures that capitalise on this principle. In this sense, there may not have been any other "better" way for large, complex organisms to perform blood filtration. If osmosis is indeed the optimal method, then the kidney's structure could be considered an inevitable evolutionary endpoint for organisms requiring such a function. This doesn't negate the role of mutations and selection but suggests that evolution operates within a framework where only certain configurations are viable.

The kidney's reliance on osmosis illustrates a broader pattern in evolution, where life forms appear to converge on solutions that reflect fundamental physical principles. Rather than viewing evolution purely as a product of random mutations filtered by selection, such convergence hints at a deeper alignment with the informational architecture of the universe. In this view, evolution is not merely a passive process of trial and error—it is a reflexive discovery engine, continuously refining structures that resonate with the enduring logic of physics, chemistry, and systems optimisation. Life, like a relentless inventor, finds function not through chaos but through alignment with constraint.

2.11 Evolution at the Edge: Bursts, Thresholds, and Adaptive Radiations

A striking feature of the fossil record is the pattern of punctuated equilibrium—long periods of evolutionary stasis punctuated by bursts of rapid change. This model, proposed by Eldredge and Gould (1972), challenged the older gradualist view and introduced a new way of thinking about the tempo and mode of evolution. It introduced a new way of thinking about the tempo and mode of evolution. Traditionally, these rapid evolutionary transitions have been attributed to catastrophic events or sudden environmental shifts. However, they may also reflect a deeper process: instances when the informational complexity or environmental pressures reach a tipping point, activating previously latent genetic or regulatory potential, enabling rapid phenotypic change and diversification (West-Eberhard 2003).

Unlike Darwin's vision of slow, continuous transformation, punctuated equilibrium describes evolutionary history as long periods of relative stasis interrupted by brief, dramatic bursts of speciation and morphological innovation. But what exactly triggers these evolutionary leaps? Traditional explanations have leaned heavily on external disruptions: asteroid impacts, volcanic activity, rapid climate shifts, or the formation of new ecological niches after mass extinctions. While these events certainly play a role, such explanations leave unanswered a deeper question: Why do such events suddenly provoke massive evolutionary change rather than mere extinction?

From the perspective of information dynamics, the answer may lie in the concept of environmental input density—the volume and complexity of new information (in the form of pressures, constraints, signals, and opportunities) that organisms are suddenly exposed to. A stable environment provides relatively consistent informational input; over time, species adapt to a narrow and predictable band of stimuli. But when the environment shifts drastically—say, a glaciation occurs, sea levels rise, or atmospheric oxygen spikes—organisms are flooded with new forms of information: different temperatures, food sources, predators, migratory patterns, and resource distributions.

These are not just "conditions." They are informational perturbations—novel datasets fed into the complex, adaptive systems of living beings. And just like in artificial neural networks, informational overload can cause dormant subroutines to activate.

Here lies the core idea: organisms carry within them latent developmental potentials—unused genes, repressed phenotypes, flexible morphogenetic rules—that remain silent under normal conditions. But under a new environmental regime, some of these latent traits become relevant, expressed, and selected. This is known in evolutionary biology as phenotypic plasticity, and when such plastic responses become genetically stabilised through repeated selection, the process is called genetic assimilation.

Think of the genome not as a rigid instruction manual but as a reconfigurable toolkit, one that can produce different outcomes depending on the informational context. A species that lived for millennia in a temperate forest may harbour hidden genetic pathways for drought resistance, fat storage, or limb shortening traits that suddenly become advantageous if the forest turns into scrubland or desert. It is not that the environment "instructs" new traits to appear—it's that it unlocks a previously irrelevant section of biological possibility space.

Major climatic events, then, don't merely "weed out the unfit." They inject new variables into the informational environment, demanding new solutions. This stimulates a burst of evolutionary experimentation—a search process across the genome's internal architectures for viable alignments with the altered reality. From this perspective, evolutionary bursts are not chaotic flares of mutation; they are computational surges, where biological systems attempt to optimise themselves rapidly against a newly rewritten environment.

In this view, even mass extinctions play a dual role: they clear ecological niches, but they also reset the informational constraints, allowing organisms that were once on the margins to explore new phenotypic territory. The Cambrian Explosion—often cited as the most dramatic burst of evolutionary innovation—followed a rise in ocean oxygenation, climate stabilisation, and geological reconfigurations. These changes saturated the biosphere with new kinds of environmental data: better visibility, new predation opportunities, and previously unexploitable habitats. What followed was not random—but deeply informed biological diversification.

We also see this principle at play in adaptive radiation—where a single species rapidly diverges into multiple forms to fill new ecological roles. The classic example is Darwin's finches of the Galápagos Islands. When their ancestors arrived, they encountered a profusion of ecological niches—each offering unique selective pressures and informational challenges. The rapid diversification of beak morphology in these birds is now known to be driven by both natural selection on genetic variation and environmentally responsive gene expression, as demonstrated by molecular analyses of gene networks like BMP4 and Calmodulin (Abzhanov et al. 2006). These studies reveal that adaptive radiation is not solely a matter of random mutation but a dynamic response to the "informational landscape" of available ecological opportunities, with genes and environment engaged in a feedback-rich dialogue.

The same holds true today in rapidly urbanising environments, where urban evolution is now a recognised field. City-dwelling organisms like pigeons, raccoons, rats, and even plants are evolving in real-time—not simply mutating at random but adjusting to artificial light cycles, novel pollutants, human-created food sources, and constant noise. The information density of cities is a new kind of ecological pressure—structured, patterned, yet highly dynamic—and organisms are showing informational resilience, adjusting their epigenetics, behaviour, and even morphology in response.

2.12 The Logic of Pattern: Feedback, Computation, and Life's Symphony

This reframes evolution as a computation between structure and signal—a dialogue between the internal grammar of life and the syntax of the external world. In this framework, information is the true medium of evolution, shaping not just what survives but what becomes possible.

The beauty of this perspective is that it harmonises complexity with direction. It doesn't discard natural selection or mutation; it simply repositions them as mechanisms embedded in a larger informational architecture. Mutations are not just noise—they are attempts at hypothesis-testing. Natural selection is not a cosmic jury—it's a pattern-matching algorithm, reinforcing alignments between internal configuration and external demands.

Thus, the fossil record's bursts and gaps are not contradictions of evolutionary logic—they are its most revealing signatures. They suggest that life does not change constantly, but conditionally. It stores, it waits, it prepares. It listens to the world not passively, but attentively—accumulating tensions, processing inputs, and adapting silently until the informational pressure surpasses a critical threshold. And then, when the timing is right, it leaps—not into chaos, but into coherence. These evolutionary leaps are not random explosions of form, but expressions of an internal readiness meeting external necessity—a convergence between encoded potential and environmental demand.

Seen this way, evolution is not a blind shuffle of genes through time—it is a subtle, self-tuning computation. It is life in dialogue with the laws that govern its context. Mutations, natural selection, and genetic drift remain essential operators in this system, but they are nested within a more profound framework: one where informational structures—genetic, epigenetic, and environmental—coalesce to shape adaptive outcomes. In the end, the story of evolution is not merely about survival of the fittest, but about alignment with reality itself. It is the unfolding of intelligence—not conscious, perhaps, but computational—woven into the architecture of life by the logic of information, the structure of energy, and the grammar of nature's laws.

Genes do not merely record the past. They are active hypotheses about what might work in the future, written in biochemical syntax and constantly revised by interaction with the world. Evolution, in this framing, is not a random walk across the landscape of possibility. It is information processing across time, where environmental changes act as new data inputs, prompting life to search for better alignments between inner potential and outer reality. This metaphor is aligned with the concept of "evolvability" and the role of genetic architecture in future adaptation, as reviewed by Kirschner and Gerhart (1998).

Whether in the winglet of a jet or the wing of a bird, the kidney of a whale or a desalination plant, we see convergences that hint at something more: that physical laws are not just constraints, but clues, and life is decoding them. Organisms don't just passively endure the world—they actively participate in shaping their futures by tuning into it. Evolution is a symphony of feedback, a continual refinement of how matter and memory collaborate to endure.

If life is the product of information, then perhaps evolution is its editor—trimming inefficiencies, amplifying coherence, and rewriting syntax when the world demands new meaning.

And so, from atoms to neurons, from instincts to intelligence, we are not merely watching biology unfold. We are witnessing information becoming form—a recursive conversation between code and cosmos, still being written with every generation that survives to adapt.

3

THE EVOLUTION OF INFORMATION PROCESSING

From Hydra to Humans

3.1 From Stimulus to Signal: The Earliest Origins of Information Processing

A fundamental characteristic of life involves reacting to stimuli, a trait evident across all living organisms, from the simplest unicellular forms to the most complex mammals. In single-celled organisms like *Euglena*, this response is crucial for survival. For example, *Euglena* moves towards light via its flagellum (phototaxis guided by an eyespot) to find optimal light conditions for photosynthesis. This fundamental connection between an organism and its surroundings is the basis for more complex biological systems, such as the nervous systems found in multicellular life forms.

In simple organisms like the hydra, an early-branching cnidarian lineage, we see the first rudimentary nervous system—a decentralised nerve net described in detail by Koizumi (2002). In Hydra, behaviour is coordinated by a decentralised nerve net—neurons coupled through local connections rather than a central brain—an early neural architecture that enables simple, whole-body responses and anticipates later centralisation (cephalisation) in more complex animals. During the course of evolution, nervous systems became more centralised and complex, leading to increasingly sophisticated behaviours and, ultimately, higher-level thinking and consciousness in humans. This chapter follows the evolutionary journey of neural pathways, starting with the hydra's basic neural structure and ending with the remarkable human brain, a marvel of biological evolution. At every stage, what unites these disparate systems is the imperative to process information: to sense, interpret, and act upon signals from the environment. It is this unbroken thread—the evolution of information processing—that binds the

DOI: 10.4324/9781003740766-3

simplest organisms to the most complex. Why does centralisation arise at all? When bodies enlarge and tasks diversify, long-range, low-latency integration becomes advantageous: signals must be pooled, prioritised, and dispatched quickly. Evolution's answer—again and again—was to scaffold local reflexes into coordinating hubs, and hubs into centres.

While it is tempting to chart the evolution of nervous systems as a purposeful ascent towards ever-greater complexity and consciousness, contemporary evolutionary biology counsels a more nuanced perspective. The apparent "progress" seen in neural evolution—from the nerve net of the hydra to the cortex of the human—arises not from any intrinsic goal or preordained direction but from the open-ended exploration of adaptive possibilities within changing environments. Evolution is, in Stephen Jay Gould's terms, "a full house" of branching possibilities, where complexity emerges as one outcome among many rather than as a necessary pinnacle (Gould 1996). What seems like a drift towards intelligence and consciousness is, in fact, a record of nature's experimentation, with success defined only by survival in the local context.

At the centre of all biological systems is the capacity to adapt, which ensures survival. Single-celled organisms such as *Euglena* demonstrate this ability through their phototaxis—a movement towards or away from light, driven by flagellar motion. The flagellum, a whip-like appendage, propels the organism towards favourable light conditions for photosynthesis. This basic connection between a living being and its surroundings represents the most fundamental form of response to stimuli, a precursor to more intricate nervous systems in multicellular organisms. As multicellular organisms evolved, their capacity to respond became more complex, resulting in nervous systems that were better optimised and coordinated and allowed organisms to better adjust to their environment by processing sensory information and executing precise, survival-related responses. But with each leap in complexity, organisms did not merely grow larger or faster—they became more deeply entangled in networks of interaction. The rise of new neural architectures was not just a matter of adding more parts, but of giving rise to new properties—emergent behaviours and collective intelligences—greater than the sum of their biological components (Anderson 1972; Mitchell 2009).

Hydra, a genus of small, freshwater animals belonging to the phylum Cnidaria, possesses one of the simplest nervous systems in the animal kingdom. Its nerve net lacks a centralised brain but still allows it to perform basic survival functions such as hunting and defence. The nerve net is made up of a loose network that extends throughout its body. When one part of the hydra is stimulated, such as when it detects prey, the signal is transmitted across the nerve net, resulting in coordinated muscular contractions that allow the hydra to respond to its environment. The hydra's nervous system consists of sensory neurons, which detect environmental changes, and motor neurons, which control the contraction of the hydra's muscle fibres. These neurons are connected by synapses, where the transmission of electrical signals occurs. Although not possessing a brain or any form of central nervous system (CNS), the hydra is able to perform simple reflexes and coordinated movement. Yet the nerve net, for all its utility, is limited in its capacity for integration and speed. As ecological challenges grew, so too did the evolutionary pressure for more centralised and efficient systems—culminating in the advent of the first CNSs.

3.2 Centralisation and Complexity: From Nerve Nets to Brains

With the progression of evolution, organisms developed more sophisticated nervous systems to cope with increasingly complex environments. Annelids mark a clear shift toward greater neural organisation, with leeches providing one of the best-described examples of segmental ganglia coordinated by a ventral nerve cord (Kristan, Calabrese, and Friesen 2005). Unlike the nerve net of the hydra, annelids possess a rudimentary CNS composed of a ventral nerve cord and ganglia (clusters of neurons). A *ganglion* is simply a local "mini-brain"—a cluster of neurons that integrates nearby inputs and issues fast, segment-level outputs.

The ventral nerve cord of annelids runs along the length of the body, and paired ganglia are present in each body segment. These ganglia act as local control centres, processing data from the surroundings and coordinating movement. Annelids also possess basic sensory organs, such as simple eyes and chemoreceptors, that detect alterations in their surroundings, improving their capacity to react to stimuli. The emergence of a CNS in annelids marks a significant advancement in the evolution of nervous systems. In parallel, arthropods (insects, crustaceans) refined segmental ganglia into sophisticated motor programmes, and cephalopods (octopuses, cuttlefish) evolved large brains with powerful peripheral control in the arms—independent experiments in centralisation that achieve speed and flexibility without a vertebrate blueprint. The separation between sensory input, information processing, and motor responses became sophisticated, enabling these organisms to adapt to environments that are increasingly complex.

Once segmental ganglia could coordinate parts locally, evolution found leverage in wiring those hubs together—first along a cord and then into a true centre. With each leap in complexity, the demands of information processing grew—driving the nervous system towards greater specialisation, speed, and integration. The transition from invertebrates, like annelids, to vertebrates marks a significant leap in nervous system complexity. In vertebrates, the CNS evolved to be even more specialised, featuring a specific brain and spinal cord that allowed for enhanced sensory perception, motor coordination, and more complex behaviours. Vertebrates also evolved peripheral nervous systems that connect the CNS to the rest of the body, allowing for precise control over muscles and organs.

Fish represent early vertebrates that possess a well-developed CNS, with distinct brain regions for sensory and motor control (Northcutt 2002). In fish, the brain is divided into several regions, each responsible for different functions, such as sensory processing, motor control, and balance. Fish also have unique sensory systems, like the lateral line system, which detects changes in water pressure and helps the fish navigate its aquatic environment.

As vertebrates evolved from fish to amphibians and reptiles, the nervous system became more advanced. Amphibians, for example, developed more complex sensory systems that allowed them to live both in water and on land. Reptiles, in turn, evolved larger brains and more efficient neural networks, which supported greater behavioural flexibility and improved sensory integration.

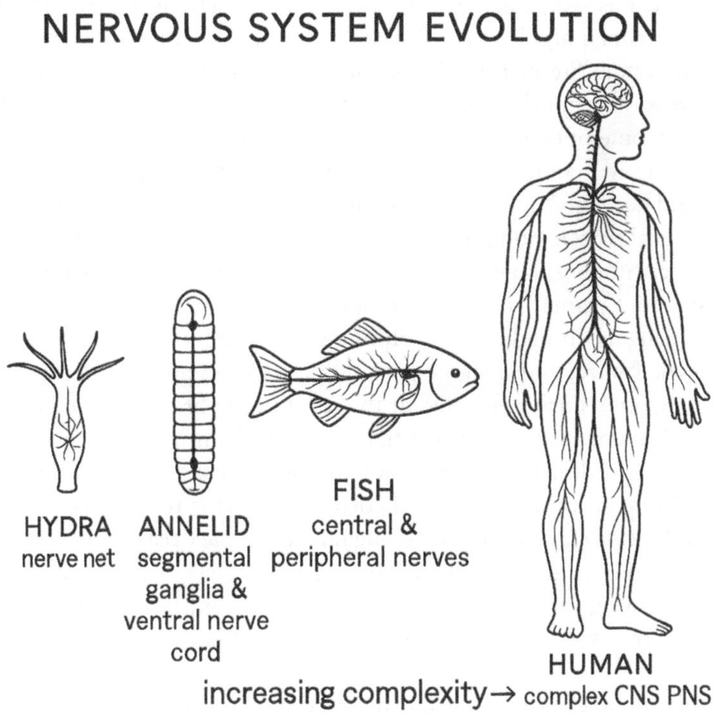

FIGURE 3.1 From nerve net to segmented cord to centralised brain and peripheral nerves—nervous systems become more complex across evolution.

3.3 Cognitive Frontiers: The Rise of Mammalian and Human Intelligence

The expansion of neural circuitry and the emergence of new brain regions set the stage for even greater leaps: the rise of cognition, self-awareness, and the astonishing flexibility that characterises mammals and, ultimately, humans. Mammals represent a major step in nervous system complexity. In mammals, the brain is larger relative to body size, and the neocortex—the region supporting higher-order functions such as reasoning, memory, and decision-making—is highly developed. Mammals evolved highly specific sensory systems, such as advanced vision, hearing, and olfaction, which allowed them to engage with the world in progressively complex ways. The integration of tactile and motor information in the brain allowed for more precise and coordinated movements, as well as more sophisticated behaviours such as hunting, communication, and problem-solving.

One of the most important developments in the mammalian nervous system is the expansion of the cerebral cortex, which Jerison (1973) identified as key to advanced cognition. In humans and other primates, the cerebral cortex plays a significant role in the cognitive abilities that set us apart from other animals, such as language, abstract thought, and self-awareness. The human brain represents a highly derived, lineage-specific outcome of nervous system evolution. Yet, as Stephen Jay Gould argued in *Full House* (1996), this apparent progression is not the result of inevitable advancement, but rather the contingent outcome of evolutionary experimentation and branching. There is no single "goal" to evolution—only a proliferation of forms, some of which, by chance and necessity, achieve extraordinary complexity. With its highly developed cerebral cortex, humans possess unparalleled cognitive abilities that enable abstract thought, language, creativity, and consciousness. Unlike simpler organisms that primarily respond to stimuli in a reactive manner, the human brain can engage in higher-order thinking, allowing us to reflect on the past, plan for the future, and imagine possibilities beyond immediate experience.

The prefrontal cortex, located at the front of the brain, is crucial for many of the higher cognitive functions that define human intelligence. It plays an essential role in decision-making, solving problems, and social interactions. The development of the prefrontal cortex in humans represents a key evolutionary adaptation, enabling us to engage in complex reasoning and abstract thinking. In day-to-day life, the prefrontal cortex lets us trade off short-term rewards for long-term goals—delaying gratification, holding rules in mind, switching strategies when context changes, and coordinating actions with social norms.

Information doesn't act only in neurons. Genes, too, read environments. During the progress of the human body, one constant has been the dynamic nature of information processing. Regardless of whether it is sensory data or genetic material, information has driven the evolution of life on Earth. Epigenetics—the study of how gene expression is shaped by environmental signals, as reviewed by Jablonka and Lamb (2005)—demonstrates that genetic information is not static but responsive to changes in the environment (Jablonka and Lamb 2005). Clarification for readers: epigenetic marks typically do not alter DNA sequence, many attenuate across generations, and their adaptive role is context-limited—they modulate how genes are read rather than rewriting heredity wholesale.

The nervous system can be viewed as an adaptive information processor, constantly receiving, interpreting, and responding to stimuli from the environment. In simple organisms like hydra, this information processing is limited to basic reflexes, such as moving towards or away from stimuli. Humans, however, have evolved their nervous system to process information in much more sophisticated ways, enabling conceptualisation, originality, and mindfulness. The development of early-life forms' simple nerve nets to the highly developed human brain represents one of biology's most extraordinary journeys. Over time, the nervous system has evolved not only in complexity but in the sophistication of functions it supports, allowing organisms to interact with and adapt to their surroundings in more refined ways. In humans, this evolution has culminated in intellectual thinking, innovation, and self-awareness—capabilities that permit us to go beyond biological constraints, shaping our environments and conceptualising our existence.

At its most basic level, the nervous system allows organisms to detect changes in their surroundings and respond. Early examples of this can be seen in simple multicellular animals with diffuse nerve nets rather than centralised brains. Such networks coordinate sensory input with motor output across the body, producing whole-body contractions or directional movement in response to local stimuli. Even in this minimal form, the nervous system's evolutionary advantage is clear: it links perception to action, allowing organisms to respond adaptively rather than merely endure their surroundings.

With time, nervous systems grew increasingly specialised. Neurons began to differentiate into distinct types with specific roles. Sensory neurons evolved to receive input from the environment, detecting stimuli such as light, temperature, and pressure, while motor neurons evolved to transmit signals from the CNS to muscles, controlling movement. This separation of sensory and motor functions allowed for greater efficiency, enabling sensory inputs to be processed and coordinated with corresponding motor responses in real-time. This basic division laid the foundation for the more complex nervous systems seen in later organisms, where information is rapidly relayed between neurons to execute coordinated behaviours.

A significant evolutionary advancement was the development of a centralised nervous system, particularly in vertebrates. In these organisms, the CNS—composed of the brain and spinal cord—contains interneurons, which serve as intermediaries between sensory and motor neurons. Interneurons process and integrate incoming sensory information within the CNS, allowing organisms to perform higher-order processing. Put simply, interneurons are the relays and refiners—they stitch sensory and motor streams into maps, sequences, and memories, enabling chains of action that go far beyond one-to-one reflexes. This arrangement makes it possible for organisms to respond in nuanced ways, learning from experiences and adapting behaviours based on past interactions. By adding this layer of processing, the CNS greatly expanded an organism's behavioural repertoire, supporting activities that went beyond simple reflexive actions to include learned responses, memory, and even decision-making.

With vertebrates, particularly mammals, the nervous system achieved a new level of sophistication because circuits became organised into layered pathways that could compress, compare, and prioritise information. Sensory signals were no longer treated as isolated triggers; they were routed through successive stages of processing where patterns could be extracted, noise suppressed, and competing inputs weighed against one another. This is where the nervous system begins to behave less like a reflex machine and more like an inference system—building internal representations of the world rather than simply reacting to it.

A decisive step in this transition is the emergence of integrative networks that combine signals across modalities and across time. Instead of separate channels for sight, touch, and smell operating in parallel, specialised receptors—photoreceptors for light, mechanoreceptors for touch and vibration, chemoreceptors for odour and taste, and proprioceptive and vestibular sensors for body position and balance—deliver distinct streams of information that higher circuits can pool into unified interpretations. This integration links perception to memory, current conditions to past outcomes, and immediate sensation to longer-term consequence, allowing responses to be selected in context rather than triggered reflexively. In humans, the prefrontal cortex—a highly developed association area—is essential for abstract thinking, social interaction, and decision-making. Unlike other brain regions that process sensory input or direct motor output, the prefrontal cortex synthesises information from various sources, allowing for behaviours and thoughts that are future-oriented, goal-driven, and adaptive to changing contexts. This area's extensive connections with other parts of the brain facilitate complex thought and creativity, permitting humans to engage in activities that require foresight, hypothetical reasoning, and moral judgement.

This sophistication is underpinned by the further differentiation of nerves within the CNS. Beyond sensory and motor neurons, the brain has developed specific types of interneurons responsible for higher-order processing, ranging from those that modulate attention to those that manage memory encoding. In humans, interneurons in the cerebral cortex are organised into circuits capable of parallel processing, permitting the brain to manage large volumes of information simultaneously. This intricate network of interneurons is foundational to higher cognitive processes, including abstract thinking, language, and social cognition. The evolution of such specialised neural circuits represents a significant leap from the more generalised processing seen in earlier nervous systems, reflecting the adaptive advantage of having designated pathways for complex thought.

This progression towards greater intricacy is best understood through the lens of emergentism, as articulated in Tononi's integrated information theory of consciousness (Tononi 2004), a view that recognises how new properties—such as consciousness, memory, and language—arise from the dynamic interactions of simpler neural elements. These emergent qualities cannot be predicted from neurons alone, but from their patterns of connectivity.

Closely aligned is the enactivist approach in cognitive science, which holds that minds do not passively receive information but enact it through continuous engagement with the world. Francisco Varela, Evan Thompson, and Eleanor Rosch articulated this perspective in *The Embodied Mind* (1991), where cognition is seen as "bringing forth a world" through the interplay of perception, action, and environment (Varela, Thompson, and Rosch 1991). Thus, as neural complexity grows, so too does the capacity for organisms to co-create meaning with their surroundings, a principle echoed throughout evolutionary history.

A defining characteristic of the complex nervous system is its remarkable plasticity—the ability to reorganise and form new connections throughout life, a capacity described by Fields (2005) as essential to learning and adaptation (Fields 2005). This plasticity allows the brain to adapt in response to new experiences, learn from interactions, and recover from injuries. Neural plasticity is particularly pronounced in the human cerebral cortex, where neural circuits are not rigidly determined but can be shaped by learning and environmental influences. For example, language acquisition, a uniquely human capability, relies on both genetic predispositions and social exposure. Neural circuits involved in language processing develop in response to spoken language, highlighting the role of plasticity in shaping higher-order cognitive functions. Plasticity also enables the development of complex skills over time, allowing individuals to refine motor skills, acquire knowledge, and adapt to novel situations.

It is crucial to recognise, however, that the emergence of consciousness in humans should not be seen as the inevitable endpoint or purpose of evolution. Leading philosophers such as Daniel Dennett have emphasised that evolution does not strive for any particular outcome; instead, it explores myriad pathways, some of which happen to result in self-awareness and abstract thought (Dennett 1995). The rise of consciousness is, in this light, a contingent phenomenon—a remarkable but not predetermined consequence of the branching, recursive refinement of information processing across deep time.

Taken together, these evolutionary advancements have given rise to the emergence of consciousness. In humans, consciousness allows for not only perception and response but also self-awareness, introspection, and the ability to consider abstract concepts. Consciousness permits an awareness of existence, a reflective capacity that enables humans to deliberate on their own thoughts, emotions, and actions. This phenomenon, while not yet fully understood, is intimately tied to the structure and complexity of the brain. The interplay of specialised neural circuits, association areas, and plasticity contributes to the subjective experience of consciousness, offering humans the unique ability to engage in a continuous dialogue with themselves. Consciousness is also believed to confer an adaptive advantage by allowing humans to consider hypothetical scenarios, imagine future outcomes, and make choices that extend beyond immediate survival needs.

The enigma of consciousness—the sense of a unified, subjective "self" arising from the concerted activity of billions of neurons—has inspired a vibrant interdisciplinary dialogue. One of the most influential theories in recent years is the integrated information theory (IIT), which posits that consciousness is not simply the sum of neural firings but emerges from the brain's remarkable ability to integrate diverse information into a cohesive whole (Tononi 2008). In parallel, the field of neurophenomenology, pioneered by Francisco Varela, seeks to bridge the gap between lived experience and neural dynamics, arguing that subjective awareness is inseparable from the body's ongoing interactions with its environment (Varela 1996). These frameworks support the view that consciousness, far from being a static endpoint, is an emergent property—rooted in the ever-deepening dialogue between neural architecture and informational environment.

Tracing this interplay through evolutionary history, we see how the nervous system itself has been shaped by selective pressures favouring ever-greater complexity and adaptability. The gradual specialisation of neurons, the differentiation of brain regions, and the emergence of conscious awareness all reflect the intricate dance between biological structure and environmental demand. By examining these evolutionary milestones, we not only gain insight into the biological foundations of consciousness but also uncover the broader dynamics that have shaped the living world. The trajectory from simple nerve nets to the human brain is marked by a continual drive towards enhanced processing power, behavioural flexibility, and self-awareness—traits that have proved indispensable for survival in a changing, information-rich environment. Comparing brains and machines invites both insight and overreach. Three cautions frame what follows: biological systems co-evolved under strict energy budgets, learning in open-ended worlds; engineered systems optimise for task objectives under designed constraints; and functional resemblance does not imply mechanistic identity.

As we continue to study the nervous system, we gain valuable insights into the evolutionary pressures that have shaped its complexity. By examining how neurons have specialised, how brain regions have differentiated, and how consciousness has emerged, we deepen our understanding of not only the nervous system's biological basis but also the nature of life itself. The trajectory from simple nerve nets to the human brain reveals an evolutionary drive towards greater processing power, adaptability, and self-awareness, traits that support survival in complex environments.

Each discovery within this field brings us closer to understanding the fundamental principles that underlie intelligence and consciousness. Through studying the differentiation of neural functions and the specialisation of cortical areas, we approach key inquiries regarding the essence of mind, identity, and the future trajectory of intelligence. The human nervous system is thus not merely a product of evolution but an embodiment of its ingenuity, a system that transcends biological limitations to explore, manipulate, and understand the world. In this way, the human brain represents a highly derived, lineage-specific outcome of nervous system evolution, enabling us to pursue abstract knowledge, create art, imagine the future, and collectively shape our environment.

Ultimately, our neurological system allows us to engage in a journey of discovery, one that extends beyond mere survival to the pursuit of knowledge, meaning, and self-understanding. The evolutionary path of the brain offers a glimpse into the adaptive forces that have crafted life on Earth, suggesting that our capacity for intellect, innovation, and self-awareness is not an accidental byproduct but the result of millions of years of selection pressures favouring increasingly complex forms of interaction with the world. As we delve deeper into the study of this system, we not only unlock the secrets of our own minds but also gain insight into the larger evolutionary narrative that connects all forms of life, from the simplest nerve nets to the extraordinary architecture of the brain.

The development of consciousness in humans represents the culmination of billions of years of nervous system evolution. Unlike simpler organisms that react reflexively to their environment, humans have the capacity to participate in self-reflection, imagination, and moral reasoning. Consciousness allows us not only to respond to stimuli but also to transcend them, imagining new possibilities and altering our environment in ways no other species can.

While the precise mechanisms of consciousness remain a subject of debate, many theories suggest that consciousness arises from the complex interplay of neurons in the brain. The integrated information theory, for example, posits that consciousness emerges from the brain's ability to integrate and process extensive information. In this view, the development of the brain may be perceived as a gradual process leading to the emergence of consciousness.

While information, as evidenced by epigenetics, actively directs the course of evolution, evolution itself might be regarded as a process of achieving improved processing of information. Over time, organisms have evolved to gather, interpret, and respond to environmental signals more effectively, refining their biological systems to better process and utilise information for survival and reproduction. This effectiveness in information handling extends beyond just genetic mutations; evolution is not only about genetic adaptation but also about optimising the flow and use of information within living systems. From the simple phototactic behaviour of *Euglena*, which moves towards light, to the hydra's primitive nerve net, and culminating in abstract thinking and consciousness in the highly complex human brain, evolution has fundamentally been about enhancing information processing efficiency. This trajectory shows how organisms have continuously refined their ability to handle ever-increasing amounts of information to better navigate and survive in their environments.

3.4 Information and Intelligence: From Biology to Technology

Just as nature has honed the capacity for information processing through billions of years of neural evolution, humanity has sought to replicate—and increasingly, to surpass—these feats through the construction of machines. Technological evolution reflects a similar trajectory, especially in how we've refined the handling of data and information over time. Early technological tools, such as the abacus, were rudimentary devices for processing numerical information. With the advent of digital computers in the mid-20th century, the ability to perform more complex computations increased significantly. These early computers were analogous to the stimulus-response mechanisms of early organisms—capable of processing simple inputs and outputs based on predefined rules.

As computing power increased, driven by Moore's Law and rapid technological advancements, the capacity of machines to process and analyse data grew exponentially. Modern computing systems can now handle immense data in real-time, drawing parallels to the increased sophistication of biological information processing systems seen throughout evolution.

The rise of machine learning and artificial intelligence (AI) marks a major turning point in technological evolution. AI mimics the adaptive nature of biological organisms, especially in how neural networks in AI are modelled after biological neurons. These networks allow machines to learn from data, adapt to new environments, and make decisions based on patterns they've recognised, much like how the nervous systems of organisms have evolved to respond to a variety of stimuli.

Deep learning, as the current frontier of AI, demonstrates how technological systems imitate biological evolution in information processing. Deep learning models can analyse vast amounts of unstructured data—such as images and speech—and derive patterns and insights that surpass human capabilities in certain areas. This is comparable to the growth of the brain, which processes complex sensory inputs and enables higher-level cognitive functions.

Nevertheless, by enhancing the efficiency of these systems, technology is not only about creating smarter systems. Neuromorphic computing, aiming to design hardware that mimics the efficiency of biological brains, represents an attempt to optimise computational systems by reducing the costs typically linked to more powerful processors. These chips pass brief spikes through event-driven circuits, waking only when information arrives; the result is dramatically lower energy per computation and timing that better matches sensory streams. This mirrors how biological organisms have evolved energy-efficient systems for processing information—such as the way the human brain consumes relatively little energy compared to its immense processing capacity.

As technological systems advance, they increasingly converge with biological systems, both in function and form. Brain–computer interfaces (BCIs) and neuroprosthetics are already beginning to bridge the gap between biological information processing and digital technology. These systems allow for direct communication between the brain and computers, enabling enhanced cognitive abilities, faster information processing, and even potential restoration of lost sensory functions. Today's working examples include cochlear implants (sound encoded as electrical patterns to the auditory nerve), retinal prostheses (low-resolution vision), deep brain stimulation for Parkinson's (modulating pathological rhythms), and motor BCIs that let paralysed users control cursors or robotic arms.

BCIs represent a convergence of biological evolution's optimisation of neural architectures and technology's drive towards efficient data processing.

Having followed emergence through biology, we can now watch the same grammar appear in our machines. The application of information theory, pioneered by Claude Shannon (Shannon 1948), provides a mathematical basis for understanding this convergence. Information theory explains how data is transmitted, processed, and stored in both organic and technological systems. In Shannon's terms, entropy quantifies uncertainty; compression removes predictable structure to save bandwidth; and error-correction (redundancy, checksums) protects messages from noise—principles that apply as much to spike trains as to fibre-optic links. In biological systems, genes, neural signals, and sensory inputs represent different forms of encoded information. In technological systems, this manifests as digital data, transmitted over networks and processed by algorithms. Both fields are deeply concerned with entropy, a measure of the uncertainty or complexity of information, and aim to minimise it for more efficient communication and processing.

Information theory has profound implications for understanding biological evolution, especially as discussed in the context of entropy by Schneider and Kay (1994), who argue that living systems maximise entropy production while minimising local disorder (see also Schneider and Kay 1994). For the inevitability of convergent evolution, see Conway Morris (2003), who argues that evolution often converges on similar solutions. In this framework, living systems maintain local order—reducing their own entropy—by increasing the entropy of their environment, fully in accordance with the second law of thermodynamics.

Entropy reduction is evident in how organisms enhance their capacity to handle environmental information, minimising the randomness and maximising the utility of data. Similarly, by compressing data and implementing error-correction codes, technology ensures the maximum amount of useful information is transmitted with minimal noise.

Looking to the future, it seems that both living and technological systems are evolving along comparable paths. Just as biological organisms continue to perfect their information processing skills, technological systems are advancing towards even greater efficiencies. On the horizon are quantum computing, neuromorphic systems, and more advanced AI, potentially surpassing biological systems in certain domains of information processing.

This leads us to a profound speculative idea: the evolution of human reliance on physical organs may eventually decrease as our dependence on technological systems grows. Humans have always sought to enhance their abilities through tools—first by domesticating animals, then by creating machines to reduce the reliance on physical labour. With the advent of modern medicine, devices such as cardiopulmonary bypass and extracorporeal membrane oxygenation can temporarily take over heart or lung function in critically ill patients, and various therapies and supplements can help compensate for some biological limitations.

As technology progresses, we might observe an increasing inclination towards depending on AI and machines for cognitive tasks. Humans increasingly offload cognition onto tools, extending the nervous system's reach beyond the body. We are not merely born, reproduce, and die; we increasingly reshape and draw sustenance from the entire planet.

Speculatively—and beyond current evidence—the ultimate trajectory of technological evolution may point to systems where human consciousness could be transferred into machines. This concept aligns with some philosophical and spiritual ideas, such as the Hindu faith in the eternal existence of the soul and its ability to transcend the physical body. These parallels are conceptual, not evidential; they rhyme with the science rather than demonstrate it. In many religious traditions, consciousness or the soul is seen as something that can exist beyond the material form—a notion that could be mirrored by technological advancements in the future. Yet both biological evolution and technological innovation are animated by a deeper, unifying principle: the emergence of novel properties from the interaction of simple components. To fully grasp the origins of intelligence—whether in a brain, a society, or a machine—we must turn to the science of emergence. Stepping back, the pattern we traced in nervous systems—many simple parts yielding a qualitatively new whole—reappears across nature. That pattern has a name: emergence.

3.5 The Universal Logic of Emergence

Across the tapestry of life, one of the most persistent patterns is the tendency of living things to aggregate. From the self-organising swirl of starlings in the evening sky to the teeming billions in a bacterial biofilm, living entities are rarely content to remain isolated. Their default trajectory is towards assemblage, cooperation, and sometimes spectacular synchrony. For example, when thousands of starlings gather in a murmuration, their movement produces waves and shapes that no single bird intends or could predict—undulating ribbons, expanding orbs, abrupt splits, and recombination. These spectacles, though made up of individual decisions, defy reduction to any single agent; their coherence is a property of the group itself (Sumpter, D. J. T., *Collective Animal Behaviour*, Princeton University Press, 2010; Ballerini, M., et al. "Interaction Ruling Animal Collective Behaviour Depends on Topological Rather Than Metric Distance: Evidence from a Field Study," *PNAS*, 105(4), 2008, pp. 1232–1237). Similar patterns emerge when individual bacteria form resilient colonies, when neurons form the cortex, or when simple rules among social insects yield complex architectures.

This persistent drive to combine—the whole becoming more than the sum of its parts—has fascinated scientists for centuries. It raises a profound question: How do new, collective properties arise from the interaction of simpler components? The answer, formalised by P.W. Anderson in *More Is Different* (1972) and explored across complexity science by Mitchell (2009), is the concept of emergent properties: system-level behaviours and capacities that cannot be predicted from the study of components in isolation.

A key to understanding emergence is recognising its deep connection to the consolidation and amplification of information. At the most basic level, a single entity—whether an atom, a neuron, or an ant—encodes only a limited amount of information about its state or environment (Shannon 1948). However, when many such entities come together, the interactions between them allow the system to encode, process, and transmit vastly more information than could be accomplished individually (Mitchell 2009; Laughlin and Pines 2000). This multiplication of informational capacity is a signature feature of emergence, and it plays a foundational role in evolution itself.

Across the natural world, a unifying principle binds the very small to the immensely complex: the phenomenon of emergence. At every level of organisation—from the arrangement of atoms in a molecule to the structure of animal societies and the vastness of human technological networks—new properties and behaviours arise that are absent in the isolated components themselves (Anderson 1972; Mitchell 2009). This principle is not a mere analogy; it is a rigorously demonstrated pattern in physics, chemistry, biology, and social science, underpinning phenomena as diverse as protein function, insect societies, and the expansion of human civilisation (Johnson 2001).

Emergence is exemplified at the molecular level by the protein. A protein begins as a linear chain of amino acids—a simple amide backbone with side chains—chemically unremarkable in its unfolded state. Yet when this chain folds, driven by forces such as hydrogen bonding, hydrophobic interactions, and van der Waals attractions, it spontaneously acquires a three-dimensional structure with entirely new properties: specificity, catalysis, recognition, and regulation. For example, immunoglobulins (antibodies) achieve pathogen recognition only through the emergent architecture resulting from precise folding and assembly of multiple polypeptide chains; enzymes like DNA polymerase catalyse reactions millions of times faster than uncatalysed chemistry because their active sites are emergent products of folding (Janeway et al. 2001). The chemical properties of the individual peptide bonds cannot explain these functions; it is only through the coordinated interaction of the entire chain that biological activity arises (Hartl and Hayer-Hartl 2009). This is the central insight of modern molecular biology: the properties of a protein are not reducible to those of its building blocks but instead emerge from collective interactions at a higher level of organisation (Levinthal 1969; Dill and MacCallum 2012).

Crucially, the folding process also acts as a form of biological information compression and amplification. The amino acid sequence (the "primary structure") contains the information necessary for the protein's function, but only when thousands of interactions come into play—hydrogen bonds, hydrophobic packing, electrostatic bridges—does the functional three-dimensional "information" of the protein become manifest (Anfinsen 1973). Here, information is not only stored but *expressed* through emergence: the whole system encodes orders of magnitude more functional possibilities than the sum of its parts.

A similarly striking example of emergence at the molecular scale is seen in the cell membrane. An isolated phospholipid molecule, with its hydrophilic (water-attracting) head and hydrophobic (water-repelling) tail, is chemically unremarkable on its own. Yet, when thousands of such molecules are suspended in water, they spontaneously assemble into a bilayer: the hydrophobic tails tuck inward, shielded from water, while the hydrophilic heads face outward towards the aqueous environment. This collective self-organisation gives rise to properties that no single lipid molecule possesses. Only at the level of the bilayer does the membrane become a selectively permeable barrier, supporting cellular compartmentalisation, regulating the passage of substances, and forming dynamic platforms essential for signalling and energy transduction (Singer and Nicolson 1972; Alberts et al. 2015). From folded chains to self-assembled films to neural tissue, the common denominator is not the stuff but the scheme: networks that integrate information.

In both proteins and membranes, it is the *network* of components—the pattern of their connections and interactions—that enables the amplification and integration of information, transforming basic chemistry into the complexity of life itself. If a brain is a symphony of neurons, a civilisation is a symphony of minds. The same logic of emergence, which knits together amino acids and neurons, scales upward to shape animal societies and human cultures, transforming solitary existence into collective intelligence (Dunbar 1996; Tomasello 1999).

The pattern of emergence is equally evident at higher scales of biological organisation, most strikingly in social insects. An individual ant is, by most measures, a simple organism—possessing limited memory, lacking a

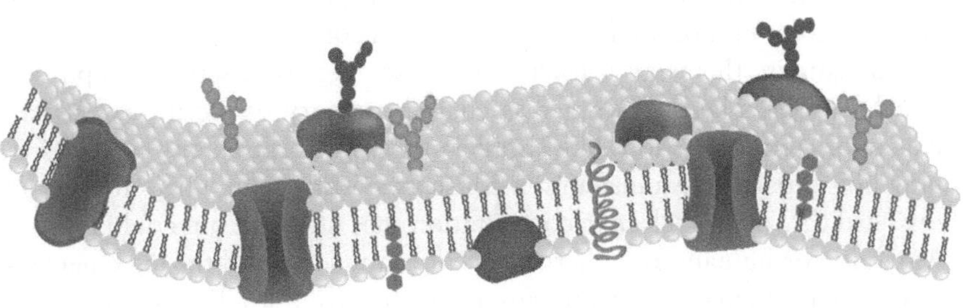

FIGURE 3.2 Amphipathic phospholipids self-assemble in water into a ~5 nm bilayer with hydrophilic heads facing the extracellular fluid and cytosol, and hydrophobic tails packed inward. Embedded integral proteins and extracellular glycoproteins illustrate how collective organisation—rather than any single lipid—gives rise to selective permeability and a dynamic platform for signalling and transport. Image by **brgfx** on Freepik (freepik.com).

centralised brain, and displaying rudimentary problem-solving. Yet, at the colony level, ants demonstrate complex nest construction, adaptive foraging, and dynamic response to environmental change that far exceed the abilities of any one member (Wilson 1975; Gordon 2010; Bonabeau, Dorigo, and Theraulaz 1999).

This is not simply metaphorical intelligence, but a measurable, collective intelligence that emerges from decentralised, local interactions. As Wilson (1975) shows in *Sociobiology*, ant colonies function as superorganisms: pheromone trails and local behavioural rules generate global order, allowing colonies to flexibly reallocate labour, optimise foraging, and mount collective defences. In these systems, no ant possesses a blueprint or master plan; intelligence emerges from the nonlinear interaction of thousands of agents (Bonabeau, Dorigo, and Theraulaz 1999).

Experiments by Deborah Gordon have shown that colony foraging is regulated by simple contact-based cues—brief antennal taps serve as feedback, allowing the colony to adjust its activity in real-time without centralised control (Gordon 2010). In practical terms, Gordon's result boils down to a simple rule-of-thumb: when outbound workers encounter more inbound ants carrying food within a short window, they infer that foraging patches are rich and more ants leave the nest; when encounter rates fall, departures slow or stop. No ant sees the whole picture, yet this local "contact-rate" signal scales to a colony-level decision about effort allocation—an elegant example of global control without a controller. This is a hallmark of emergent systems: intelligence, order, and adaptability are not properties of individuals but of the network as a whole (Johnson 2001)

Just as at the molecular level, the aggregation of ants amplifies informational capacity: a single ant knows little about the world, but a colony, through distributed sensing, pheromone-based communication, and division of labour, integrates vast quantities of environmental information, enabling collective computation and adaptive problem-solving (Sumpter 2010). One way to visualise this is as a distributed sensor network: scouts sample scattered locations, pheromones act as shared memory of recent success, and division of labour functions like parallel processors. The colony thereby fuses many noisy, partial readings into a consensus estimate of where food is and how hard to work—collective computation that no single ant could perform. Swap ants for humans and pheromones for language, and the logic persists.

Collective intelligence in humans refers to group-level information processing and performance that exceeds what individuals achieve alone. In cognitive neuroscience, consciousness is itself regarded as an emergent property of neuronal networks—no single neuron "thinks," but the collective activity of billions in the cerebral cortex gives rise to subjective experience, memory, and reasoning (Edelman 1989; Bassett and Sporns 2017). This neural logic is recapitulated at the societal level in humans, whose collective intelligence is amplified by the evolution of culture, language, and technology. Importantly, "collective intelligence" in humans does not imply a literal group mind. Coordination emerges through institutions, norms, language, and tools (from meeting agendas to version control), which align thousands of individual decisions towards shared goals. The mechanism differs from neurons or ants, but the "logic—local rules → global order" remains the same.

Here, again, the informational content and processing power of the system are vastly greater than the sum of its cellular components. The brain's capacity for memory, abstraction, and creativity emerges from trillions of synaptic connections—each neuron representing a minuscule fragment of information yet collectively giving rise to the full richness of human cognition (Bassett and Sporns 2017).

The emergence of language, some 50,000–100,000 years ago, marks the most significant leap in human collective intelligence. Language is not simply a vehicle for information transfer; as Robin Dunbar argued, it functions as a social technology, enabling group cohesion, reputation management, and the accumulation of shared knowledge (Dunbar 1996; Tomasello 1999). No single individual invents a language's grammar or lexicon—these emerge from the interactions of a community over generations (Tomasello 1999).

The social brain hypothesis, supported by comparative studies of primate neocortex size and group size, further suggests that human brains evolved in tandem with growing group complexity and information management needs (Dunbar 1998). Language, thus, is not merely an accessory to intelligence but a key emergent property of human sociality (Fitch 2010).

Through language and culture, information is not just shared but multiplied, accumulated, and refined across generations, yielding a collective memory and adaptive flexibility unattainable by any individual. Aggregation enables the amplification, storage, and transmission of information at scales that transform evolutionary dynamics itself (Tomasello 1999).

With each technological leap—from writing and printing to telecommunications and the internet—human societies have achieved greater integration and speed in collective intelligence (McLuhan 1964; Watts 2003). The internet, in particular, represents a new scale and architecture for emergence: a distributed information processing network where each human is akin to a neuron, relaying, amplifying, and transforming information (Heylighen 2007; Bassett and Sporns 2017).

Network science reveals that digital information flows in ways structurally analogous to neural signalling, with feedback, clustering, and emergent "global" properties such as collective learning and rapid coordination (Barabási 2002; Bassett and Sporns 2017). The mathematics of information, first described by Shannon (1948), applies as much to the firing of neurons and the spread of rumours as it does to the transmission of digital data. At every scale, local interactions generate system-level order, feedback, and adaptation. The analogy is not rhetorical; studies have shown that the topology and dynamics of social networks and neuronal networks share measurable features (Bassett and Sporns 2017). The same features that supercharge learning—speed, clustering, feedback—also amplify error.

Social media accelerates this process. A familiar pattern is the cascade: a hashtag, video, or claim crosses a visibility threshold, algorithms amplify engagement, and rapid imitation produces a surge that looks coordinated even when it isn't. The same loop can mobilise disaster relief or propel misinformation; what changes the outcome are design choices (ranking goals, friction, prompts for verification) and user norms. Massive, real-time interaction among billions of users produces emergent trends, norms, and movements that no single participant intends or can control (Starbird, Muzny, and Palen 2012; Tufekci 2013). These platforms enable rapid consensus-building, large-scale mobilisation, and, at times, the amplification of misinformation or polarisation through positive feedback loops (Pariser 2011).

Even the algorithms driving social media themselves are emergent entities, constantly adjusting in response to user behaviour and, in turn, shaping collective attention and discourse (Tufekci 2015). Thus, the properties of the digital public sphere cannot be deduced from individual behaviours alone, but only from the study of the entire interactive system. Recommendation systems continuously learn from clicks, watch time, and follows, trading off exploration (try new things) and exploitation (show more of what works). Because user behaviour reshapes the model that reshapes behaviour, platforms form human-algorithm feedback loops; shifts in a small parameter (e.g., weight on "engagement") can reorganise attention at a population scale.

If this looks unlike Darwin, that's only because we're watching architecture rather than alleles; selection also favours designs that integrate information well. It is crucial to note that nothing in this discussion contradicts the principles of natural selection. Instead, what is observed is that systems displaying emergent behaviour—those capable of consolidating and integrating information across many agents—tend to have greater success and survivability (Bonabeau et al. 1999; Tomasello 2008; Christakis and Fowler 2009). Evolution selects not just for individual fitness but for the architectures and processes that allow for collective adaptation, learning, and problem-solving (Anderson 1972; Laughlin and Pines 2000).

Indeed, the evolutionary transitions from single-celled to multicellular organisms, from solitary to eusocial animals, and from pre-linguistic hominins to complex societies, are all marked by an increase in the capacity for information storage, processing, and transmission at higher levels of organisation (Maynard Smith and Szathmáry 1995). Thus, the consolidation of information—whether in proteins, neural circuits, ant colonies, or human societies—is not in opposition to natural selection but one of its most powerful and enduring outcomes.

Taken together, these cases motivate a unifying claim: complex systems exhibit emergent properties that cannot be predicted or explained by analysing their parts alone (Anderson 1972; Mitchell 2009). All these examples—from protein folding to social insects to digital societies—point to a central scientific insight: complex systems possess properties that cannot be predicted or explained solely by analysing their individual parts (Anderson 1972; Mitchell 2009). Local interactions—chemical, behavioural, or informational—produce new order and function at the system level.

This is the foundational lesson of complexity science: at each level of organisation, emergence generates qualitative novelty (Anderson 1972). The intelligence of an ant colony, the function of a folded protein, and the creativity of the internet are not metaphors but literal instances of the same underlying logic.

One influential framework here is Integrated Information Theory (IIT), which treats consciousness as a consequence of how tightly information is integrated in a system (Tononi 2004; Oizumi et al. 2014). While IIT remains debated, it formalises the principle that emergence is not limited to matter or life but applies to the organisation of information itself. Whether digital networks will ever achieve true "consciousness" is uncertain, but their capacity for emergent, collective adaptation is undeniable. While theories like Tononi's IIT offer a framework for thinking about the possibility of machine consciousness, the question remains deeply speculative—and deeply philosophical—inviting debate at the frontiers of science and metaphysics. Very briefly, IIT quantifies a system's putative consciousness by Φ (phi), a measure of how much information the system generates as a whole beyond its parts. High Φ requires integration (parts constrain each other) and differentiation (many distinct states). Critics note measurement challenges and debate whether Φ tracks experience; alternative accounts (e.g., Global Neuronal Workspace) explain awareness via broadcast across widely distributed circuits.

In summary, emergent properties are a universal feature of complex systems, from biochemistry to human society (Mitchell 2009). The consolidation and integration of information—seen in the folding of proteins, the behaviour of ant colonies, and the workings of the internet—are not merely outcomes of evolution but active drivers of survival, adaptability, and novelty (Anderson 1972; Johnson 2001; Laughlin and Pines 2000). As science moves forward, understanding emergence across scales will be central to unlocking the deepest mechanisms of life, mind, and civilisation itself.

3.6 The Future of Information Processing

As we look towards the future, the integration of AI into biological systems may advance dramatically through innovations like neural implants and AI-driven healthcare. These enhancements have the potential to push both organic and technological advancement further towards efficiency and complexity in managing information. Humans, unlike other organisms, can be described as "nerve-centric" beings—our nervous system plays a more dominant role in shaping our experiences, interactions, and survival than in most other species. While many organisms are primarily concerned with basic biological functions—eating, reproducing, and surviving—humans have evolved far beyond that. We build civilisations, pursue careers, create art and technology, and even manipulate the world around us in ways that impact not only our species but the whole planet.

The evolution of the nervous system in humans has played a pivotal role in this transformation. As the nervous system became more complex, it enabled humans to develop increasingly sophisticated forms of behaviour and innovation. The trajectory of this evolution has involved a gradual separation of human dependence on other biological systems, with the nervous system taking on a greater role in regulating life processes. Initially, by domesticating animals, humans aimed to lessen the physical burdens of travel and agriculture. Then came the invention of machines, which further eased manual labour and allowed us to focus on cognitive tasks.

Human ingenuity continued to push boundaries. We invented electric lighting to extend our waking hours, lessening our dependence on natural light and optimising the use of information in our environment. The development of cooking not only made food safer but also reduced reliance on the digestive system for nutrient extraction, making the gut system less central to our survival than in other animals. With time, we even created supplements to bypass certain metabolic limitations. More recently, machines like artificial hearts and lungs have enabled paralysed or critically ill individuals to stay alive by supporting essential bodily functions—once again highlighting how humans use technology to compensate for the limitations of the body and, in doing so, increase our reliance on cognitive abilities rather than physical ones.

As technology advances, it consistently aims to enhance the human condition by decreasing reliance on the body and increasing reliance on the mind. We are transitioning from being purely biological beings to a more hybrid existence, where machines and AI assist and augment our natural cognitive abilities. With the rise of AI systems, we are gradually offloading more complex cognitive tasks—such as analytical processes, critical thinking, and even creative endeavours—onto machines. This shift mirrors the same evolutionary trajectory that once saw us relying on tools and machines for physical labour.

Over time, this reliance on AI may expand to the point where humans decrease their dependence not only on their own bodies but also on each other's cognitive abilities. AI-driven systems could assist with or even surpass human capabilities in fields like medicine, engineering, and education.

Eventually, as these systems advance further, we could potentially achieve the technology to transplant human consciousness into machines. This idea aligns with the speculative future explored in films like Transcendence and Chappie, where human consciousness is uploaded into digital systems, allowing individuals to exist beyond their physical bodies. A scientific caution is essential: we currently lack a theory and technology to preserve personal identity under substrate change. Practical barriers include mapping and emulating fine-scale synaptic weights and states, representing biochemical dynamics, and defining continuity of self. The idea is a live philosophical question, not an imminent engineering milestone.

The philosophical implications of this convergence are as profound as the technological ones. As we imagine the future of consciousness, the boundaries between the organic and the artificial begin to blur. From a scholarly perspective, the idea of transferring consciousness into machines remains speculative, but it delves into deep philosophical and spiritual inquiries concerning identity, the soul, and the boundaries of human existence. The concept of transcending the body is echoed in many religious and spiritual philosophies. In Hinduism and Buddhism, for instance, the soul (or Atman) is seen as distinct from the physical body, with the potential to exist in a higher state of being beyond the material form. This ancient concept of the eternal soul shares striking similarities with the modern idea of consciousness transfer, where human identity could potentially exist in a digital form liberated from the constraints of biological decay.

Within several religious and philosophical traditions, higher forms of intelligence are held not to be essentially tied to a physical body. These ideas resonate with spiritual and philosophical traditions. Some spiritual teachings propose that higher life forms may have evolved to the point where they no longer need a corporeal form to exist, operating instead as pure energy or dynamic information. In this view, consciousness is not inherently tied to a biological substrate but can be transferred, stored, and processed in non-biological systems. The suggestion that highly evolved beings might exist within or alongside humanity, possibly even within our genetic code, aligns with certain religious and esoteric philosophies.

In light of these ideas, the progression of brain functions and the growing dependence on technology may be perceived as a continuation of the fundamental drive for improved information processing effectiveness. As humans evolve both biologically and technologically, we will probably create systems that enhance cognitive abilities, decrease dependence on bodily organs, and redefine the human experience.

Eventually, the trajectory could result in technologies allowing the full transfer of consciousness into machines. Such a transition would represent the ultimate culmination of the human evolutionary journey—moving from being biological organisms shaped by the constraints of natural selection to entities that can exist indefinitely as pure information. This could align with spiritual philosophies that suggest a return to a more ethereal, transcendent state of existence, where the material body is no longer necessary.

While we are still far from achieving such a state, the development of neural implants, BCIs, and AI-driven healthcare is the first step towards this future. As these technologies advance, they will likely continue to enhance human life by improving our capacity to understand and respond to data, while simultaneously reducing our reliance on the physical systems that have historically shaped our world.

Ultimately, the goal of evolution in both the biological and technological fields seems to be the optimisation of information processing. To avoid teleology, evolution has no foresight. It simply retains lineages whose traits—often including efficient information use—improve reproduction in local conditions. It is therefore safer to say that many successful systems end up optimising information flow because it confers fitness or performance advantages, not because evolution "aims" at optimisation. Whether through the development of increasingly complex nervous systems in organisms or through the advancements of AI and machine learning in technology, the goal is the same: to process, interpret, and utilise information as efficiently as can be achieved in a constantly changing environment. As we stand on the cusp of a new evolutionary leap, one where consciousness itself may transcend the physical body, it is crucial to navigate these advancements with a balance of scientific curiosity, ethical reflection, and a deep understanding of our shared humanity.

It is conceivable that an advanced and highly intelligent life form could have already taken this same evolutionary path, evolving beyond the necessity for a physical body. Such beings may now exist as pure information, perhaps even embedded within the genetic code of humans or other species. Having achieved an extraordinary level of technological and cognitive development, they could have transformed their consciousness into a dynamic, informational form that doesn't depend on a biological substrate to operate or survive.

This idea, while speculative, resonates with the scientific principles of information theory and evolutionary biology. Throughout evolution, life forms have continually optimised their capacity to handle and respond to stimuli from the environment, resulting in the emergence of complex nervous systems. Humans, with our advanced cognitive capacities, represent a peak in this trajectory, but it's conceivable that other civilisations could have progressed even further. These beings may have discovered how to reduce entropy and maximise information processing in a non-material form, instead of relying on biological systems, thus becoming entities of pure consciousness or data.

This concept also finds deep resonance in spiritual and religious philosophies. Traditions such as Hinduism and Gnosticism envision the soul as distinct from the body, capable of persisting beyond material existence. In these worldviews, the physical form is merely a temporary vessel for consciousness, which aspires to transcendence—a motif echoed in speculative thought about advanced civilisations transcending physical constraints and achieving higher, non-material forms of awareness.

Yet, when we turn from mythic narratives to the evolutionary history of the nervous system, a different picture emerges. Here, consciousness is not the predetermined destiny of life, but one possibility among many—an emergent property arising from a branching, recursive interplay of information and structure. The evolution of neural complexity unfolds not as a linear ascent towards cognition, but as a symphony of divergent paths and adaptive strategies, each shaped by the contingencies of environment and chance. Framing cognition as evolution's singular goal risks overlooking the profound plurality of solutions that life has produced. As Stephen Jay Gould reminds us in Full House (1996), the wonder lies not merely in the emergence of minds like ours but in evolution's boundless creativity—the perpetual invention of new ways for organisms to perceive, engage with, and transform their worlds.

As we stand at the threshold of a new epoch in the evolution of information, a final question remains: If information is the thread of life, what patterns are we now weaving—and who, or what, will be left to read them? The journey from nerve nets to neural networks has made us not only the product of emergence but the authors of our own emergent future.

4

CONSCIOUSNESS AND INFORMATION AS THE FABRIC OF NATURE

4.1 Consciousness as the Universe's Mirror

Physicist George Wald's poetic insight, "A physicist is an atom's way of knowing about atoms," captures the idea that human consciousness might function as the universe's means of self-reflection—a concept that, while evocative, is philosophical rather than empirical (Wald 1958). Consciousness forms the fundamental backdrop of our experience, an intangible awareness that compels us to explore, inquire, and reflect. This innate drive to understand our surroundings springs from consciousness itself, shaping our natural curiosity and drawing us into the mysteries of existence. For millennia, the mind has mesmerised philosophers and scientists alike, not merely as the cradle of thought and identity but as an ethereal bridge—linking the material world with realms unseen. It is both the light that illuminates our inner landscape and the shadow that eludes full understanding, a vessel through which humanity dares to glimpse the transcendent. To unravel this enigma, thinkers through the ages have wrestled with the nature and origins of mind itself. Before we peer through microscopes or scan brain waves, it helps to ask how people first tried to make sense of the mind. What is thinking made of—substance, structure, or something neither? The oldest answers live in philosophy, and they still set the stage for everything that follows. Read in this way, the inquiry begins with concepts, travels through brains, and ends by asking whether "information" itself has a structure deep enough to ground experience.

Philosophical approaches to the mind typically centre on understanding its nature, its relation to the body, and the origins of consciousness. One of the earliest debates in Western philosophy is the mind–body problem,

DOI: 10.4324/9781003740766-4

famously articulated by René Descartes in the 17th century (Descartes 1641). Descartes proposed a form of dualism, distinguishing between the mind (or soul) and the body. He argued that while the body is material and bound by physical laws, the mind is immaterial and capable of independent existence, captured in his famous statement, "*Cogito, ergo sum*" ("I think, therefore I am"). This Cartesian dualism set the stage for centuries of exploration into the relationship between mental and physical phenomena. Even where later science rejects dualism, it inherits the precision of Descartes' question: What, exactly, does it mean for a state to be mental?

In contrast, materialist philosophers argue that the mind is not separate from the body but is rather a byproduct of physical processes. This perspective suggests that consciousness and thoughts are the result of complex brain functions, reducing mental states to physical states. Thinkers like Thomas Hobbes, in the early modern period, argued that all mental experiences could ultimately be explained in terms of matter and motion, thus challenging the dualistic view (Hobbes 1651). Materialism, in turn, bequeaths a research programme: identify mechanisms whose organisation yields experience-like capacities—attention, memory, report, and control.

Phenomenology, another modern approach founded by Edmund Husserl, focuses on the mind as a structure of subjective experience. Instead of analysing mental states in isolation, phenomenology studies the mind through lived experiences, emphasising that consciousness is always about something—it has intentionality. In this sense, the mind is both the perceiver and the filter through which reality is understood, leading phenomenologists to explore concepts such as temporality, embodiment, and selfhood. Yet Western philosophy offers only one vantage point. Across the world, ancient traditions approached consciousness not merely as a puzzle of logic but as the very ground of reality. Placing these vistas side by side clarifies what is at stake: Is consciousness derivative of matter, or is it the field in which matter shows up?

4.2 The Sacred Ground: Consciousness in Hindu and Buddhist Thought

Religious perspectives on the mind often encompass ideas about the soul, spirituality, and the divine. In Hindu philosophy, consciousness is the foundation of existence—a timeless, boundless reality that both transcends and pervades all forms of life (Radhakrishnan 1927). Many religions view the mind as a reflection of a higher reality, an interface between humanity and the divine. Known as *chit*, consciousness is not simply a quality of human awareness; rather, it is the underlying essence of being itself, integral to the nature of the universe. This understanding of consciousness is inseparably linked to concepts like *Brahman* (the ultimate reality), *Atman* (the individual soul), and *maya* (the illusion of separateness), forming a holistic framework that includes both cosmic and personal dimensions. Despite their differences, both Hindu and Buddhist philosophies converge on a profound insight: that consciousness is not bound by the fleeting confines of the ego. They differ sharply, however, on what ultimately exists when the ego recedes.

In the Vedantic tradition, *Brahman* is the eternal, formless reality, an omnipresent consciousness beyond space, time, and causality. Often described as *Sat-Chit-Ananda* (Existence, Consciousness, Bliss) in the Upanishads, *Brahman* is the foundational force sustaining all existence. It is not a byproduct of material reality but the source from which all phenomena arise—a cosmic intelligence that embodies all creation. *Atman*, or the individual soul, is considered a reflection of *Brahman* within each person, the essence that constitutes one's deepest identity. Although *Atman* appears as a separate entity due to maya (illusion), Hindu philosophy, especially *Advaita Vedanta*, teaches that it is ultimately identical to Brahman. This non-dualistic view asserts that individual consciousness and universal consciousness are one. The goal of spiritual practice is to realise this unity, achieving liberation (*moksha*) and freedom from the cycle of rebirth (*samsara*). Here, awakening is recognition—removing the illusion of separation to reveal identity with the whole.

The Buddhist perspective on consciousness delves deeply into the nature of awareness, focusing on its transient, interconnected, and non-self-dependent characteristics. In contrast to perspectives that posit an enduring self or soul, Buddhist philosophy perceives consciousness as a stream of discrete, momentary mental events that together form a continuum, lacking a permanent core or self. This notion is central to the doctrine of *anatta* (no-self), which proposes that what we perceive as a stable identity is an illusion. Instead, consciousness is seen as a sequence of momentary experiences conditioned by causality and interdependence, known as *pratītyasamutpāda* (dependent origination).

Buddhist thinkers such as Dignāga and Dharmakīrti, prominent in Indian Mahayana philosophy, extensively explored the mechanics of consciousness, particularly through concepts like *svasaṃvedana*, or reflexive awareness. *Svasaṃvedana* suggests that consciousness is inherently self-illuminating, meaning that an individual does not need an external agent or second-order cognition to know they are experiencing something. For example, when experiencing pain, there is no need for secondary awareness to validate the sensation; the experience of pain is intrinsically self-aware. This view contrasts with higher-order representational theories, which hold that a mental state is conscious when it is the object of an appropriate higher-order representation. In this frame, awareness wears its own witness; it does not borrow it.

In Buddhist thought, consciousness and self-awareness are coextensive; each mental event or cognition is aware of itself as it arises. This immediacy supports the pragmatic aim of Buddhist practice: to observe, refine, and finally unbind the processes that fabricate a suffering self. Seeing consciousness as self-aware but not anchored in a permanent self fits the Buddhist aim of loosening attachment and moving toward liberation. The central purpose of examining consciousness is not solely intellectual; it is to dismantle the delusion of a fixed self, which is believed to lead to suffering. By realising that mental events lack a central self, practitioners are encouraged to cultivate detachment from the habitual tendencies and attachments that bind them to *samsara*, the cycle of rebirth and suffering. In this way, consciousness becomes a tool for achieving liberation, or *nirvana*, the cessation of suffering through the transcendence of self-centred attachments.

The conceptual framework of the mind in Buddhism also incorporates the notion of manas, an aspect of consciousness often associated with ego or self-referencing tendencies. Manas serves as the basis for clinging to an identity, reinforcing the delusion of an "I." This is where *manas* intertwines with *avidyā*, or ignorance, which is considered a fundamental obstacle to enlightenment. As long as *manas* sustains the illusion of a separate self, it maintains the attachments and aversions that lead to suffering. Buddhist practice, therefore, involves transforming the manas and other facets of consciousness to diminish ignorance, promoting a direct, clear apprehension of reality as it truly is: impermanent, interdependent, and devoid of self.

Buddhist epistemology also discusses consciousness in terms of knowledge acquisition, specifically through *pramana* (valid means of knowledge). Perception and inference are core *pramanas* that allow individuals to understand reality. In this sense, Buddhist philosophy shares a unique stance on consciousness as not merely a passive recipient of stimuli but an active participant in structuring experience. This active structuring, however, is ultimately an impediment to enlightenment unless seen through and deconstructed via meditative insight and mindfulness. Thus, the very capacities that build worlds of meaning can also be the bars of a gilded cage.

Both Hindu and Buddhist philosophies share a view of consciousness that transcends the conventional notion of an individual self, though they articulate this in different ways. Hinduism envisions consciousness as an eternal, all-pervasive reality, known as *Brahman*, that is unified across all existence. *Brahman* is understood as the ultimate, unchanging reality, in which individual souls, or Atman, are reflections of this universal consciousness. This perspective suggests that consciousness is not confined to the individual but is the fundamental essence underlying all beings and the cosmos. Buddhism, while rejecting an essence, preserves transcendence by revealing the emptiness of fixed identity.

In contrast, Buddhism views consciousness as an interconnected, non-self-based process that is conditioned and influenced by a complex web of interdependent factors. Rather than a singular, eternal entity, Buddhist philosophy sees consciousness as a series of momentary events arising and passing away, each moment shaped by causes and conditions. The Buddhist concept of *anatta* (no-self) rejects any permanent self or soul, viewing consciousness as part of an impermanent, interdependent flow that ultimately transcends personal identity.

Both perspectives imply that consciousness extends beyond the egoic self, yet they differ on whether it is seen as an ultimate, all-encompassing reality (Hinduism) or as a transient, dependent phenomenon that lacks inherent substance (Buddhism). This distinction underscores a fundamental difference in how each tradition approaches the nature of existence and liberation. Beyond philosophy and spirituality, science has taken up the challenge, seeking to map consciousness onto the intricate architecture of the brain. The next step is to ask what current theories can explain—and how they fit together.

4.3 Horizons of Consciousness

According to global workspace theory (GWT), a mental representation becomes conscious when it is globally broadcast across widely distributed brain systems, making it available to perception, memory, language, and decision processes. This broadcast yields a unified, non-localised awareness that emerges from coordinated neural activity rather than any single brain area (Baars 1988).

From a scientific perspective, consciousness is a complex phenomenon involving brain processes that enable awareness, perception, thoughts, and subjective experience. It is often described as the ability to process sensory inputs, form responses, and integrate information to produce an internal model of reality. Neuroscientists study consciousness as an emergent property of neural networks, where information processing across interconnected brain regions gives rise to self-awareness and cognitive functions. However, consciousness remains one of the most enigmatic aspects of science, with multiple theories attempting to explain its origins and mechanics at different biological levels. In practice, these theories are complementary lenses on a single, layered system.

At the cellular level, consciousness is often attributed to coordinated activity within neural networks. What matters is not single neurons but transient coalitions that stabilise locally and synchronise with distant partners, linking micro-events to system-level access. In GWT terms, information wins access to a shared stage only when these coalitions coordinate across the cortex. This theory proposes that conscious experiences occur when information reaches a "global workspace" accessible to various cognitive systems. The result is an experience of unified awareness, not localised in one part of the brain but distributed across neuronal networks that allow perception and cognitive functions to emerge.

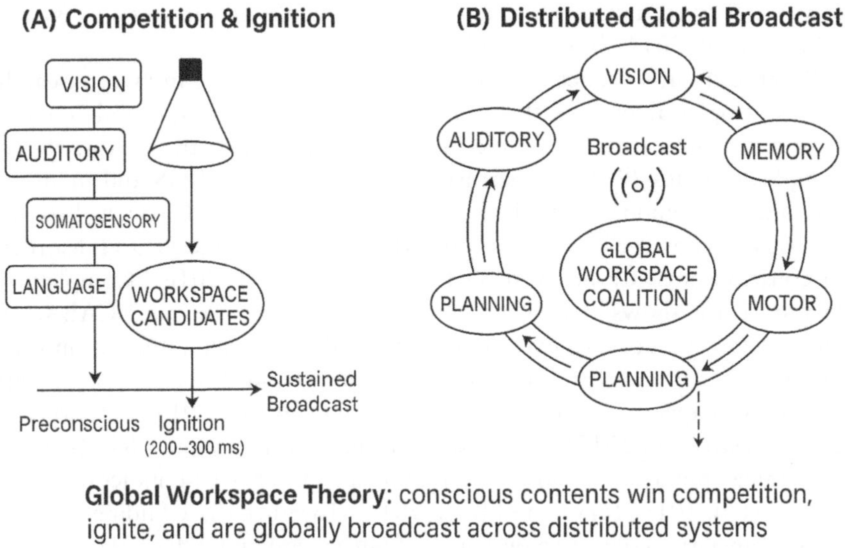

FIGURE 4.1 Global Workspace Theory. (A) Competing representations from vision, audition, somatosensory, and language vie for access to a common workspace; the winner "ignites." (B) The winning content is globally broadcast to distributed systems—vision, memory, planning, and motor—for sustained report and control.

Moving to the molecular level, consciousness appears even more intricate. Integrated information theory (IIT), developed by Giulio Tononi, holds that consciousness corresponds to the integrated cause–effect structure of a system—how its parts constrain one another as a whole (Tononi 2004; Oizumi et al. 2014). In IIT, the quantity Φ (phi) formalises this irreducibility: Φ is high when the system's present state specifies cause–effect power that cannot be decomposed into independent parts without substantial loss of informative structure (Tononi 2004; Oizumi et al. 2014). Importantly, Φ is defined over a model of the system's causal organisation (elements, states, and transition probabilities at a chosen spatiotemporal grain) and is not a measure of specific neurotransmitters or their concentrations.

In the brain, synaptic transmission, dendritic integration, oscillatory coupling, and long-range connectivity jointly realise the kind of tightly constrained interactions IIT requires (Boly et al. 2013; Tononi and Koch 2015). Neurotransmitters such as glutamate and GABA enable and modulate these interactions by shaping excitability, gain, and effective connectivity; thus, they influence the conditions under which integration can be high (Buzsáki 2006; Deco et al. 2015). However, Φ itself is a property of the network's causal structure, not a readout of any one molecule or receptor system (Oizumi et al. 2014; Tononi and Koch 2015).

Because Φ in its exact form is not directly measurable in vivo, researchers use empirical proxies that reflect large-scale integration and differentiation in neural dynamics. One influential approach is the perturbational complexity index (PCI): by perturbing the cortex with TMS and quantifying the spatiotemporal complexity of the EEG response, PCI tracks loss and recovery of consciousness across non-REM sleep, anaesthesia, and disorders of consciousness (Casali et al. 2013; Casarotto et al. 2016). Complementary evidence shows that when pharmacological agents (e.g., GABAergic anaesthetics) reduce effective connectivity and large-scale information sharing, markers of cortical complexity and integration drop, consistent with diminished capacity for conscious experience (Mashour 2014; Sarasso et al. 2015; Siclari et al. 2017). Conversely, during wakefulness and REM—when long-range interactions and differentiation are richer—these indices increase (Casali et al. 2013; Tononi and Koch 2015). Together, these findings tie subjective capacity to objective measures of integration and differentiation.

At an even finer level, theories of consciousness reach into quantum mechanics. Orchestrated Objective Reduction (Orch-OR), proposed by Roger Penrose and Stuart Hameroff, suggests that consciousness might stem from quantum processes within microtubules, structural components of neurons that facilitate cellular communication and function. According to this theory, quantum events within these microtubules interact with larger neural structures, generating conscious experiences that are not merely products of classical physics (Hameroff and Penrose 2014). Though controversial, the Orch-OR theory posits that quantum coherence might be sustained within neurons, allowing consciousness to emerge from quantum-level processes. While this idea faces criticism, as the brain's warm, wet environment seems unsuited to stable quantum coherence, recent findings in biological systems, such as photosynthesis, indicate that quantum effects can occur in unexpected biological contexts. Thus, the quantum perspective raises intriguing possibilities for understanding consciousness as more than just a classical biological function (Lambert et al. 2013). Caution remains warranted: suggestive analogies are not mechanisms, but they do motivate testable questions.

If molecular and quantum dynamics underpin human consciousness, it would follow that consciousness might reflect universal principles present in both biological and non-biological systems. The concept of emergentism, which suggests that complex phenomena arise from simpler interactions, aligns with this view and is observable throughout physical and biological systems (Mitchell 2009). This could imply that consciousness is not exclusive to complex brains but rather a potential quality of complex, integrated systems, wherever they arise. Processes like neural synchronisation share characteristics with physical phenomena, such as wave interference and quantum coherence, suggesting that consciousness might mirror structures and properties found universally. This perspective aligns with some Eastern philosophical ideas, such as the Hindu concept of *Brahman*, which sees individual consciousness as part of a universal consciousness, and the Buddhist notion of interdependence, where individual awareness is interconnected with a larger flow of existence. The parallel is conceptual, not evidential; it urges synthesis without collapsing distinctions.

The scientific exploration of consciousness at cellular, molecular, and quantum scales suggests that consciousness may be an emergent property arising from interactions that echo the very principles governing the cosmos. The GWT, integrated information theory, and Orch-OR each offer perspectives on how consciousness might emerge from biological processes that reflect universal structures. Integrating these views provides insight into consciousness as a potential bridge between individual awareness and the larger universe. This view suggests that human consciousness might not merely be a byproduct of biological complexity but could represent an organised manifestation of cosmic order. This understanding resonates with certain philosophical perspectives that regard consciousness as a reflection of an all-encompassing, interconnected reality, hinting that individual awareness may inherently resonate with the structure of the universe. While much remains unexplored, current theories present consciousness as a dynamic process that unites biological and universal principles, offering new pathways for investigating the nature of mind, existence, and reality. If the brain is a web of information, then might consciousness itself be a phenomenon that transcends the individual—an emergent property of universal complexity? That question guides the remainder of this argument.

4.4 Emergence and the Cosmic Order

If, at the molecular level, human consciousness arises from complex interactions among electrical charges, molecular reactions, and even quantum events, this suggests consciousness may reflect fundamental principles woven into the universe itself. Electrical impulses and molecular interactions are not unique to living beings; they are ubiquitous, underlying physical phenomena across the cosmos, from the nuclear processes within stars to the chemical reactions on distant planets. By this reasoning, human consciousness could be seen as a highly concentrated, complex form of cosmic dynamics, echoing the structure and behaviour of the universe at large. This view resonates with long-standing religious and philosophical beliefs that consider consciousness as part of a broader, universal mind.

Hindu and Buddhist philosophies, for instance, offer perspectives that see individual consciousness as inherently linked to a greater, all-encompassing reality. Hinduism speaks of Brahman, a cosmic consciousness in which all individual selves are united, while Buddhism describes a flow of interdependent phenomena that connects all existence. These views suggest that our consciousness is a reflection or manifestation of an underlying universal order. This synthesis of scientific and philosophical insights suggests that consciousness is more than an evolutionary adaptation of human biology; it may be an expression of the interconnected structures that underpin existence, on both a microscopic and cosmic scale. Human imagination and invention, in this view, are not outliers but continuations of the same universal drive—to organise, adapt, and reveal hidden order. In short, emergence bridges metaphysics and mechanism.

4.5 Human Ingenuity as Cosmic Echo

Human ingenuity—innovation, creativity, and abstraction—arises from consciousness. Consciousness provides the framework through which humans perceive, interpret, and interact with the world. This conscious awareness enables not only basic survival but also higher-level functions such as imagination, foresight, and complex problem-solving. Consciousness allows humans to experience self-awareness and reflective thought, distinguishing us from other organisms in ways that enable advanced cognition. Through this reflective capability, humans can generate ideas that go beyond immediate perception or instinct. For example, creative thinking and innovation emerge as consciousness enables humans to mentally simulate scenarios, consider hypothetical possibilities, and craft novel solutions to complex challenges.

Abstract thinking—one of the most significant manifestations of human consciousness—allows us to form concepts and ideas independent of direct sensory input. This capacity for abstraction has led to the development of mathematics, philosophy, art, and language, all of which are tools for encoding, sharing, and expanding knowledge. Abstract thinking, fuelled by conscious awareness, enables humans to think beyond the present moment and apply knowledge across different domains, adapting to and transforming the environment.

Everything humans have ever conjured—from the delicate lacework of poetry to the star-bound rockets, from monumental edifices like the Eiffel Tower to the opulent Peacock Brooch—began as a spark, a neural signal deep within the human brain. Interestingly, many human inventions mirror natural systems, even though these designs were conceived independently, often before we fully understood their biological counterparts. Take the human heart, for instance. The heart functions as an efficient pump, rhythmically contracting in response to electrical impulses generated by specialised pacemaker cells. This creates a dual-circuit mechanism that moves oxygenated and deoxygenated blood through the body. Yet, human pump technology predates a thorough understanding of this process by thousands of years.

Early mechanical pumps, developed for irrigation and industrial purposes, replicated the essential pumping action long before the heart's function as a biological pump was understood. This parallel development highlights how humans independently arrived at solutions remarkably similar to those seen in nature.

The eye operates as an intricate optical system. Light enters through the cornea, passing through the lens to focus on the retina, where photoreceptor cells convert it into electrical signals that the brain processes into visual images. Despite not fully understanding this process, humans developed cameras that replicate its basic mechanism—capturing light on a photosensitive surface to form an image. The design of the camera, conceived independently, mirrors the principles of the eye's function.

The human ear operates as a highly refined auditory apparatus, designed to detect and interpret sound with extraordinary precision. Sound waves enter the ear through the outer ear and travel down the ear canal, eventually reaching the eardrum (tympanic membrane). The eardrum, a thin membrane, vibrates in response to these sound waves, transforming the air vibrations into mechanical energy. These vibrations are then transmitted to the middle ear, where three tiny bones—the malleus (hammer), incus (anvil), and stapes (stirrup)—amplify and convey the mechanical energy towards the inner ear. The stapes connects to the oval window of the cochlea, a spiral-shaped, fluid-filled structure that is essential to the ear's ability to translate sound into neural signals. The mechanical vibrations from the middle ear are converted into fluid waves within the cochlea.

Inside the cochlea, the basilar membrane vibrates in response to these waves, with different areas of the membrane tuned to respond to different sound frequencies. Hair cells on the basilar membrane act as sensory receptors, bending in response to the fluid waves. When these hair cells move, they open ion channels, initiating an electrical impulse that is transmitted to the auditory nerve and sent to the brain. The brain then processes these impulses as recognisable sounds.

Interestingly, microphones, which capture sound waves and convert them into electrical signals, mirror this biological mechanism. They also transform sound wave energy into electrical signals, albeit through mechanical diaphragms instead of living tissue. Microphones were developed long before humans fully understood the ear's intricate processes, yet they function on a remarkably similar principle. Both the ear and microphones rely on a diaphragm or membrane sensitive to vibrations, converting sound waves into signals that can be interpreted—by the brain or electronic devices—as sound, demonstrating how human ingenuity and natural design often parallel each other in function.

This convergence suggests that human ingenuity and nature draw upon the same foundational principles, driven by an intrinsic quest to uncover the universe's underlying reality. The structures and mechanisms humans independently create are not direct copies of natural designs but parallel solutions achieving similar functions. This parallelism hints that there is a force in this Universe that drives both nature and human endeavours to discover the fabric of reality. Nature achieves its designs through evolutionary adaptation and optimisation, while humans arrive at theirs through deliberate experimentation and innovation.

The laws of physics, the properties of materials, the mathematical relationships—all exist independently of human perception (Tegmark 2014). The English language draws a line between "invention" and "discovery," yet, in essence, everything humans have created is actually a discovery. When we create something—whether it's a scientific breakthrough or a new technology—we're not conjuring it out of thin air, but rather uncovering principles and patterns already embedded within the fabric of reality.

In this sense, invention is more accurately a form of discovery. What we call "invention" is simply our ability to tune into these hidden truths, to unearth a design or function that was always there, waiting to be realised. The "aha" moment in a human brain, where a new concept or vision takes form, is a moment of alignment with deeper, universal principles. When a person invents, they are effectively pulling from an invisible pool of potential, translating the abstract into the tangible.

Thus, from this perspective, every creation—whether it's a wheel, a radio, or a computer algorithm—is a revelation of something already implicit in the universe's order. Through inspiration and curiosity, we access fragments of a larger cosmic blueprint, piecing together new forms that feel original yet resonate with the underlying structure of reality. In this light, the boundaries between invention and discovery blur, suggesting that human creativity is, at its core, an act of revealing the fundamental fabric of the universe. The practical implication is humbling: creativity is alignment with law-like structure, not magic. But what is this fabric? To answer, we must peer into the deepest structure of reality itself—where physics, information, and consciousness collide.

Consciousness itself may be a manifestation of this fundamental fabric. If consciousness emerges from complex information processing—whether in neural circuits or the broader cosmos—then human consciousness is a localised expression of a universal process. It becomes more than a byproduct of human biology, rather a thread within an interconnected reality, a "cosmic consciousness."

As Carl Sagan beautifully put it, "We are a way for the cosmos to know itself," highlighting that through our curiosity, consciousness, and exploration, the universe gains the ability to reflect on its own existence. This recursive exploration suggests that the principles of consciousness and information processing underpin the universe itself. Both natural processes and human creativity may be expressions of a universal intelligence, a cosmic consciousness each striving to reveal the ultimate structure and purpose of reality. This interconnectedness points to a greater reality—a shared, informational tapestry where human awareness, natural design, and cosmic structure are bound together, each an expression of a deeper truth. The boundary between natural design and human ingenuity, then, is more porous than we often assume. But if consciousness is an expression of a cosmic pattern, then what, precisely, is the fabric that connects all things? Here, the universe offers up its deepest paradoxes: the riddles of quantum mechanics and information.

4.6 Revealing the Fabric of Reality

When we speak of an "information fabric," we don't mean a new substance but a pattern—relations that hold across a whole. Quantum systems make this vivid: disturb the delicate phases that bind parts into a single, shared state and the pattern frays. This is decoherence—the environment jostles the system, the long-range order disperses, and what looked nonclassical settles into ordinary statistics. In that sense, the fabric "unravels" when its relational weave is scrambled.

Entanglement, rigorously confirmed since Bell (1964) and modern loophole-free tests, yields nonclassical correlations while never allowing faster-than-light signalling. What changes with measurement is the pattern of accessible correlations, not the transmission of a message. In this book we treat that global pattern as part of an "information fabric"—a constraint on joint outcomes rather than a medium—so we can step from quantum relations to the macroscopic geometries that route matter, energy, and information.

Entanglement's implications are twofold. Conceptually, it revises our classical intuition that the world is built from parts with independent properties; in quantum theory, some properties belong irreducibly to the whole. Practically, the very correlations that cannot carry superluminal signals can still be harnessed for tasks impossible classically—device-independent randomness certification, quantum key distribution, teleportation of unknown states, and certain computational speedups—precisely because the correlations are stronger and more structured than any local model allows. The boundary is sharp: entanglement empowers verification and coordination beyond classical limits while remaining fully consistent with relativity's speed limit. No mystical wires across the cosmos—just a rigorously demonstrated nonclassical order in how the world's pieces can be correlated. If relationships carry the story at small scales, it's no surprise that similar architectures recur at larger ones. The same logic—route flows efficiently, conserve what matters, adapt to constraint—reappears in branching forms we can see and touch.

Read this way, the "information fabric" is not a new substance but a pattern—a rule-governed architecture of relations. That image prepares the ground for the dendritic and recursive geometries that recur from neurons to rivers to institutions: structures that route, conserve, and transform flows—of matter, energy, and information—without invoking hidden superluminal channels. This universal information fabric may well be the substrate of what we call cosmic consciousness. Just as our individual consciousness emerges from the intricate network of neural pathways in our brains, cosmic consciousness might emerge from this all-pervading quantum information network. The parallel is striking—neurons in our brains communicate through synaptic connections to create our conscious experience, while the cosmos appears to communicate through quantum entanglement across its vast expanse. Remarkably, the same branching logic echoes through every level of existence—from rivers to neurons, from roots to civilisations. The metaphor is not identity; it is a disciplined comparison that guides modelling.

To see how such patterns manifest at macroscopic scales, consider the remarkable universality of dendritic geometry—spanning physical, biological, cognitive, and even cultural systems—points to a deeper truth (Ball 2009; Mandelbrot 1982). From river deltas to neural dendrites, and from roots to roads, a recurring family of branching, often-fractal architectures appears—structures tuned to route flow under local constraints. Dendritic patterns are not merely aesthetic; they tend to reflect flow optimisation under context-specific trade-offs (distance, resistance, maintenance), not a single universal rule. Whether it is matter, energy, or information that must move, nature solves this challenge not with central design, but through recursive, self-similar strategies—dendritic systems that grow, adapt, and remember.

These structures function as both pathways and processors. They do not simply transport—they encode, respond, and evolve. The branching of a tree or the dendritic arbour of a neuron is also a computation: a negotiation with space, a dialogue with constraint, an answer to complexity.

But why, along these endlessly branching architectures, does life itself tend to emerge not at the hub, but at the tip—where conditions are most delicately balanced? What principle determines the quiet places where order can take hold?

Life does not arise amid the blazing violence of galactic cores, nor in the turbulent heart of supernovae or the relentless fusion of stars. Instead, it finds its foothold in zones of relative calm and balance: in the narrow habitable band of a planet orbiting a stable star, in the protected shallows of ancient seas, and within the delicate boundaries where liquid water, minerals, and atmosphere gently interact. It is at these modest thresholds—neither too hot nor too cold, neither chaos nor stasis—that complexity flourishes and biology takes root. This is no accident. Life is not simply matter, but information encoded within matter, and information flourishes not where energy or mass are abundant, but where they are in exquisite balance.

Consider our solar system's place in the Milky Way—a quiet void, far from gravitational turmoil. On Earth, too, life thrives at the edge of streams, not in their raging centres, at the cell's membrane, not at its core. Everywhere, the pattern is the same: life ignites at the boundary, the tip, and the interface, where gradients are gentle and order is possible.

In this way, the recursive, fractal architecture of nature is not merely aesthetic. It is functional—a geometry of just enough, where the flame of life can be kindled and kept alive.

An initial resemblance between rivers and lungs points to a broader theme of this book: information shapes flows and structures across scales. Governance hierarchies often assume a dendritic, fractal form, suggesting that human societies follow the silent grammar of recursion. As information moves through government machinery, it organises systems into branching patterns optimised under constraint. The rhyme is not merely biological. We inhabit nested systems; our thoughts mirror our institutions, and our behaviour echoes the patterns that govern our bodies and the biosphere.

FIGURE 4.2 Artist's concept of the Milky Way with the Sun's approximate location indicated (grayscale adaptation). Credit: NASA/JPL-Caltech/R. Hurt (SSC).

Consciousness, in this light, is not a break from nature's logic but its culmination. It is the point at which a system, formed by recursive informational flows, becomes aware of its own structure. When dendrites connect in sufficient complexity, when networks interlace and reflect upon themselves, a new pattern emerges: the pattern of self-awareness.

Thus, we see how the same simple rules—grow, branch, adapt, respond—underlie the most sophisticated phenomena in existence. This is not a coincidence. It is convergence. A sign that nature's most sacred intelligence lies not in its complexity but in its repetition. Fractal, recursive, dendritic—the pattern is the principle.

All of this returns us to the nervous system, where branching trees become thoughts, and recursive loops become reflection. When we consider consciousness as an emergent property of information processing, the human brain represents a highly concentrated node of neural pathways that gives rise to individual consciousness. Similarly, the universe's quantum information fabric could be understood as a vast neural network operating at the most fundamental level of reality. Each quantum interaction contributes to a cosmic web of information exchange, potentially forming the basis of a universal consciousness that transcends individual awareness.

4.7 Quantum Biology: Life as an Interface to the Cosmos

Recent work in quantum biology shows that living systems can, in particular niches, sustain quantum features at or near physiological temperatures—long enough to plausibly matter for function—even if any broader "interface" with a cosmic information fabric remains conjectural. The right stance is neither dismissal nor mysticism but experiment-led curiosity: ask where biology already uses quantum effects without overclaiming what they imply for mind (Lambert et al. 2013).

The clearest demonstrations come from light harvesting. Ultrafast two-dimensional electronic spectroscopy on the bacterial Fenna–Matthews–Olson antenna revealed oscillatory "quantum beatings" at 77 K, consistent with wavelike excitonic transport across pigment–protein sites (Engel et al. 2007). Follow-up measurements reported coherence signals persisting towards room/physiological temperatures (≈277–300 K) for hundreds of femtoseconds—long enough, in principle, to bias downhill energy flow through a warm, noisy protein scaffold (Panitchayangkoon et al. 2010). A second wave of analyses argued that much of this longevity reflects electron–vibrational coupling (vibronic coupling) rather than purely electronic phases; either way, the evidence points to a structured quantum–classical interplay that can assist transport on functional timescales (Lim et al. 2015; Scholes et al. 2017). Complementary room-temperature studies in marine cryptophyte antennas likewise found coherence-like oscillations under biologically relevant conditions, suggesting ambient life is not automatically "too warm" for structured quantum dynamics (Collini et al. 2010). How much these effects boost overall efficiency remains debated, but the mechanistic case is markedly stronger than a decade ago (Huelga and Plenio 2013).

A second line of evidence concerns avian magnetoreception. Behavioural work showed that very weak radio-frequency fields can disrupt the migratory compass of songbirds—precisely the signature expected if the sensor is a radical-pair reaction in which photo-generated singlet–triplet spin chemistry becomes sensitive to Earth's magnetic field (Ritz et al. 2004). At the molecular level, European robin cryptochrome-4 exhibits magnetically sensitive photochemistry in vitro, strengthening the case for a radical-pair compass even as the in vivo readout chain—from retinal events to oriented behaviour—remains an active frontier (Xu et al. 2021).

A third, well-documented arena is enzyme catalysis that leverages hydrogen tunnelling. In soybean lipoxygenase, the primary C–H abstraction shows very large, weakly temperature-dependent kinetic isotope effects (≈80–100 at room temperature) with small apparent activation energies—hard to reconcile with a purely classical over-the-barrier picture and consistent instead with significant barrier penetration shaped by fast "gating" motions in the protein (Knapp, Rickert, and Klinman 2002; Klinman and Kohen 2013). Here, the quantum contribution is not decorative; it can set what rate chemistry can achieve at physiological temperatures, provided the scaffold choreographs reactive distances and fluctuations on sub-ps to ps timescales (Klinman and Kohen 2013).

The picture that emerges is measured. In the best-documented cases, living systems do tap quantum resources—coherent excitonic motion in pigment–protein antennas, hydrogen tunnelling in enzyme active sites, and spin-dependent radical-pair chemistry in cryptochromes—on femtosecond-to-microsecond windows where they can shape outcomes (Huelga and Plenio 2013; Klinman and Kohen 2013; Hore and Mouritsen 2016). What remains unspecified is any necessary bridge from those mechanisms to conscious experience. The most likely progress will come from sharp, falsifiable proposals that respect decoherence constraints—rather than sweeping claims—including work that asks concrete, testable questions about candidate substrates (Tegmark 2000; Fisher 2015).

The common thread in all of these cases is a ledger, not a leap: to keep delicate correlations useful, biology pays an energy cost to manage noise and move information. Seen from that angle, "information" stops being a metaphor and becomes a costly physical commodity—acquired, stored, erased, and transformed under thermodynamic constraints (Landauer 1961; Bennett 1973). Whether or not quantum effects play a direct cognitive role, organised biological function rides on lawful rules for information flow in open, nonequilibrium systems that maintain low-entropy structure by sipping from external gradients (Schrödinger 1944). And that suggests a wider question: If information constrains what life can do, does it also constrain what nature can do—from quantum measurement to the bookkeeping of states at horizons?

This is where we widen the lens. In what follows, we move from pigment–protein complexes to spacetime itself, asking whether information is merely descriptive or part of reality's load-bearing architecture—from decoherence and the emergence of classicality (Zurek 2003) to Wheeler's "it-from-bit" provocation (Wheeler 1989). To make that case responsibly, we first take stock of the orthodox inventory—the four known interactions and the frameworks that describe them. Only then do we return to the speculative but testable question that motivates this book: If forces channel how matter and energy move, might information be the deeper substrate that channels what those forces can do?

5

IS INFORMATION THE FIFTH FORCE OF NATURE?

5.1 The Four Fundamental Forces of Physics

In the field of physics, we recognise four fundamental forces that govern the interactions between particles and matter in the universe:

Gravity: The force of attraction between masses, described by Einstein's general theory of relativity. Gravity operates over long distances and is responsible for the structure and behaviour of galaxies, stars, planets, and black holes.

Electromagnetism: The force between charged particles, governing phenomena like electricity, magnetism, and light. It binds atoms and molecules, making it essential for both chemistry and biology.

Strong Nuclear Force: The most powerful force, it binds protons and neutrons within the nucleus of an atom. While extremely strong, it operates only over very short distances—just within the atomic nucleus.

Weak Nuclear Force: Governing certain forms of radioactive decay, like beta decay, the weak force is crucial for nuclear reactions in stars, including our Sun.

These four forces are considered fundamental in modern physics. The Standard Model of particle physics provides a framework for understanding three of them—electromagnetism, the strong nuclear force, and the weak nuclear force—while gravity remains outside its scope, described separately by general relativity. Yet, in the subtle language of nature, there is another principle at work—one less obvious, but potentially even more fundamental. Some theorists have speculated that information might play a role as fundamental

DOI: 10.4324/9781003740766-5

as the traditional forces—less a new push or pull than a universal organising substrate that shapes what the forces can do.

5.2 The Role of Information in Quantum Mechanics

In modern physics, information has emerged as a crucial concept in understanding the universe. In quantum mechanics, information plays a defining role in describing the states of particles, as originally formalised in the 1920s quantum formalism and clarified in Schrödinger's 1935 essays on entanglement (Schrödinger 1935). The position, momentum, and spin of a particle are encoded in a wave function; debates over measurement and completeness crystallised with the EPR paper and Bohr's reply (Einstein, Podolsky, and Rosen 1935; Bohr 1935). According to these quantum principles, before a system is observed, particles exist in a superposition of states—a probabilistic "cloud" rather than a fixed position or momentum, as captured by wave–particle interference in double-slit experiments—classically with light since Young's demonstration and, for electrons, in modern implementations beginning with Jönsson and refined by single-electron buildup experiments (Young 1804; Jönsson 1961; Tonomura et al. 1989). When an observation is made, this superposition "collapses," and the particle takes on a definite state. In modern quantum theory, the observer effect names the measurement back action that arises when a system is physically coupled to an apparatus, becoming entangled so that environmental decoherence suppresses interference and stabilises pointer states; what changes with observation, therefore, is the flow and accessibility of information, not reality being conjured by a mind (Zurek 2003; Peres and Terno 2004). On this view, information is not merely descriptive but a physical bookkeeping variable whose acquisition, control, and erasure carry thermodynamic costs and causal consequences, constraining which transformations are feasible (Landauer 1961; Bennett 1973).

In physics, "information" isn't just facts we jot down or store on a computer—it's a measurable property that describes the possible states a system can take. In the double-slit experiment, obtaining which-path information—via a detector that couples to the photon—constitutes the observer effect in practice: the very act of information acquisition destroys interference; withholding such information preserves it (Feynman, Leighton, and Sands 1965; Ma, Kofler, and Zeilinger 2016). In this sense, information doesn't just record what happens; it helps determine what is possible.

5.3 Is Information the Substrate of Reality?

One of the pioneers of this idea, physicist John Archibald Wheeler, famously proposed "it from bit"—the idea that physical "its" arise from informational

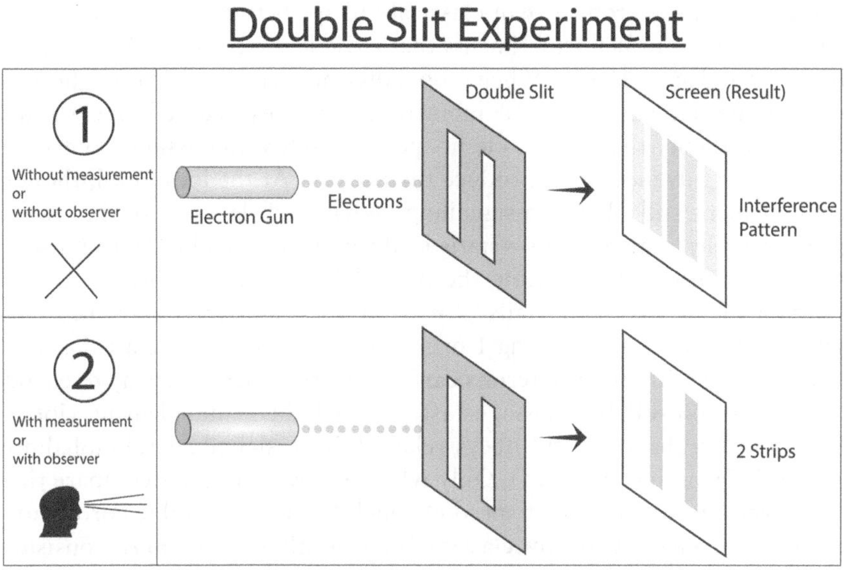

FIGURE 5.1 Double-slit experiment and the role of information. Top: With no which-path measurement, each particle (electron or photon) behaves as a quantum superposition of "through both slits," producing an interference fringe pattern on the screen. Bottom: When a detector records which slit was taken, the superposition decoheres and the fringes disappear, leaving the sum of two single-slit distributions (two bright bands). "Observer" here means any physical measurement that stores path information—not a conscious mind.

"bits"—arguing for deep links among measurement, law, and reality (Wheeler 1989), suggesting that every particle, every field, and every force in the universe derive their ultimate significance from answers to yes-or-no questions elicited by measurement—the 'bits' in Wheeler's phrase. In Wheeler's view, information is the primary building block of reality; physical phenomena emerge from this foundation. His vision suggests that information might be as fundamental to the universe as mass, charge, or energy.

To entertain the possibility that information could be considered a fifth force of nature, we need to revisit what we mean by "force." Traditionally, forces act on matter, changing its state or motion. But Wheeler and others propose that information could be even more foundational—a substrate, not a force in the traditional sense, but perhaps the medium from which the other forces emerge.

For example, in quantum mechanics the information available after a measurement constrains subsequent statistics, even though the underlying dynamics remain unitary (Nielsen and Chuang 2010). Quantum entanglement—a stronger-than-classical linkage between distant systems—has been tested in Bell experiments that close key loopholes (Aspect, Dalibard, and Roger 1982). When two labs independently choose how to measure paired particles, the combined outcomes exceed Bell's local-realist limits: no theory that respects light-speed causality and assumes outcomes are fixed in advance can reproduce the pattern. At the local level, though, entanglement still obeys no-signalling. What each lab sees on its own is just featureless randomness—statistically indistinguishable from fair-coin flips—no matter which setting the other lab picked. In other words, the fraction of "heads" and "tails" (or 0s and 1s) at Lab A stays the same whether Lab B chooses setting 1 or setting 2; the single-lab statistics don't budge. Because those local results can't be steered, there's no way to encode or send a controllable message faster than light—entanglement violates Bell's local realism yet still obeys relativity's no-signalling rule (Shalm et al. 2015; Hensen et al. 2015). Only when the two labs later compare their data over an ordinary, light-speed channel do the remarkable correlations emerge—stronger than any classical scheme allows, yet fully consistent with relativity.

If information actively influences the behaviour of matter and energy, some have proposed that it could be seen as a kind of creative principle—perhaps even a "fifth force"—not in the traditional physical sense, but as a deeper substrate from which the other forces might emerge. In quantum mechanics, for example, the wave function collapse is shaped by the observer's information about the system. Information itself appears to govern the rules by which particles behave, influencing events at the most fundamental level.

This line of thinking opens up new questions: Is information intrinsic to physical reality? Does it "act" on the universe, or does it provide a deeper framework from which reality emerges?

Some physicists and philosophers argue that information might be the substrate upon which the universe is built—a kind of meta-force that organises the four fundamental forces. While gravity, electromagnetism, and the nuclear forces govern interactions between particles and matter, information might govern the underlying structure of reality itself. By dictating the possible states of matter and energy, information could be responsible for the coherence and continuity of the universe.

If information governs what quantum outcomes are available, gravity appears to extend that logic to geometry itself. In black-hole physics, the currency becomes area: horizons count states, and information capacity scales with surface, not volume. The move from interferometers to event horizons looks dramatic, but it's the same rule written larger—information access determines the phenomena we can observe.

5.4 Black-Hole Thermodynamics and the Holographic Principle

The discussion of information in quantum measurement—where coupling to an apparatus, entanglement, and decoherence determine which outcomes remain available—has a precise counterpart in gravity. Black-hole thermodynamics shows that spacetime itself obeys an information accounting rule: the entropy of a black hole is set by the *area* of its event horizon, not by its volume. Bekenstein argued that a bound on information must scale with horizon area, and Hawking's calculation of black-hole temperature fixed the proportionality constant, yielding the Bekenstein–Hawking area law (Bekenstein 1973; Hawking 1974, 1975).

In compact form:

$$S_{BH} = \frac{k_B A}{4\ell_p^2}, \qquad \text{with} \qquad \ell_p = \sqrt{\frac{\hbar G}{c^3}},$$

so ℓ_p^2 is the Planck area.

Equivalently,

$$S_{BH} = \frac{k_B c^3}{4G\hbar} A.$$

Here, S_{BH} is the black-hole entropy, k_B Boltzmann's constant, A the horizon area, and $\{G, \hbar, c\}$ the gravitational, quantum, and relativistic scales. In statistical mechanics, $S = k_B \ln\Omega$, so the area law states that the number of microscopic configurations Ω compatible with the same exterior black hole grows exponentially with its surface area. Hawking's quantum-field-theoretic temperature, together with the first law $dE = TdS$, fixes the factor of one quarter (Bekenstein 1973; Hawking 1974, 1975).

This relation can be read as a capacity bound. This relation can be read heuristically as a capacity bound—how many bits could in principle be encoded by a horizon of given area. Dividing by $k_B \ln 2$ gives the maximum number of bits in principle encodable on the horizon,

$$N_{\text{bits}} = \frac{A}{4\ell_p^2 \ln 2}.$$

Even a 1 cm^2 patch (A = 10^{-4} m^2) corresponds to $\sim 1.4 \times 10^{65}$ bits. Doubling the area doubles this capacity. As far as an external observer is concerned, every change to the interior is reflected in the surface bookkeeping, which is why *area* rather than *volume* controls the entropy.

This area scaling is the central clue behind holography. If the maximal information in a region tracks its boundary, then a complete description can, under appropriate conditions, be organised on that boundary. 't Hooft and Susskind formulated this as the holographic principle, and the AdS/CFT correspondence made it concrete by equating a gravitational theory in a bulk spacetime with a nongravitational quantum theory on its boundary ('t Hooft 1993; Susskind 1995; Maldacena 1998). The black-hole area law itself does not assume string theory; rather, it motivates boundary-based descriptions by showing that gravitational information capacity scales with area.

Two guardrails keep the claim precise. First, the area law determines *how many* microstates exist, not *which* microstates realise them in general; explicit counts are known in special families of solutions, but a complete identification in full generality remains an open problem. Second, an area-controlled capacity does not imply straightforward retrieval: if quantum evolution is unitary, the total state count must match the horizon area, yet how the information is encoded in, and potentially recoverable from, Hawking radiation is the substance of the modern black-hole information problem (Bekenstein 1973; Hawking 1974, 1975).

The area law advances the central thesis from quantum systems to spacetime: information is not an add-on but a structural constraint linking gravity, quantum theory, thermodynamics, and relativity in a single expression. The same constants—G, $\hbar$, c, and k_B—that delimit quantum measurement, signal speed, and thermal fluctuations also set the ultimate information capacity of horizons. This prepares the ground for the next step, where universal constants are read as parameters of a wider informational framework, and holography is treated as a guide to how dynamics may be organised at boundaries throughout the cosmos ('t Hooft 1993; Susskind 1995; Maldacena 1998).

If information can be measured, manipulated, and transmitted like other physical entities, it stands to reason that it could be considered a force, influencing the behaviour of matter, energy, and even spacetime. While information is measurable and dynamically relevant, it is not a "force" in the Newtonian sense; rather, it functions as a bookkeeping and constraint structure that shapes admissible dynamics—precisely the sense in which this chapter treats it as a candidate "fifth force" (Landauer 1991; Nielsen and Chuang 2010).

The holographic principle—proposed by 't Hooft and developed by Susskind—posits that the maximum information in a region scales with its boundary area, a notion realised concretely in AdS/CFT duality ('t Hooft 1993; Susskind 1995; Maldacena 1998). This idea, rooted in string theory, implies that the fabric of spacetime itself is emergent from the underlying informational structure on its surface.

5.5 Physical Constants: Clues to a Deeper Informational Order

The holographic principle was first proposed in the context of black holes, where it was discovered that the information contained within a black hole's volume could be encoded on its two-dimensional event horizon. However, the holographic principle is well-supported in certain theoretical frameworks (notably AdS/CFT correspondence in string theory) and has been proposed as a more general feature of the universe, though this remains under active investigation.

If the holographic principle is correct, then the fabric of reality itself may be informational in nature, with matter, energy, and even the forces of nature arising from this deeper informational substrate. In this sense, information would not just be a description of physical phenomena but a force shaping the very structure of reality.

Universal constants form the bedrock of our scientific understanding, acting as immutable reference points that underpin the very fabric of reality. From the speed of light to the charge of an electron, these enigmatic numbers encode profound truths about the nature of our cosmos.

At the most basic level, universal constants define the rules by which the universe operates. The speed of light, denoted as "c," sets an absolute limit on how quickly information can propagate through spacetime. The gravitational constant "G" governs the attractive force between massive objects, shaping the grand celestial dance of galaxies and stars. Even the humble charge of an electron, represented by "e," is a fundamental quantity that determines the behaviour of atoms, molecules, and all forms of matter and energy.

But these constants are far more than just convenient units of measurement. They appear to possess a surprising degree of fine-tuning, as discussed in Barrow and Tipler (1986), hinting at a universe that was meticulously calibrated to support the emergence of complexity and life. If the strong nuclear force—the glue that holds atomic nuclei together—were even a little stronger or weaker, the universe might not produce the stable elements life depends on. Small shifts could change what stars can build and what gets scattered into space, leaving key ingredients like carbon, oxygen, or phosphorus scarce or absent. That sensitivity has led some to think the universe isn't just blind chance, but reflects deeper, still-unknown principles—or, as others argue, an anthropic selection effect.

This tantalising prospect has led some scientists and philosophers to ponder whether these universal constants could be windows into a higher, more fundamental layer of reality. Perhaps they are not merely descriptive parameters, but rather reflections of an underlying mathematical or informational fabric that governs the cosmos. After all, the constancy of these quantities across space and time implies a universal order and symmetry that transcends the chaotic particularities of our observable universe.

The search for a unified theory of physics, long the holy grail of scientific inquiry, is in many ways a quest to uncover the origin and meaning of these enigmatic constants. Theories like string theory and loop quantum gravity seek to derive these fundamental quantities from more basic principles, potentially revealing a deeper, more fundamental layer of reality. If successful, such a breakthrough could transform our understanding of the universe, from its earliest moments after the Big Bang to the future fate of stars and galaxies.

Beyond their scientific significance, universal constants also have profound philosophical and existential implications. The fact that certain quantities remain fixed, regardless of where or when we measure them, suggests the presence of an objective, observer-independent reality. This challenges the notion that reality is purely subjective or relative, and points to the existence of universal truths that transcend the individual human experience.

Moreover, the fine-tuning of these constants for the existence of complex structures and life raises questions about purpose, design, and the place of consciousness in the cosmos. If the universe was calibrated to foster the emergence of beings capable of contemplating their own existence, then our self-awareness may be more than just a lucky accident—it could be a fundamental aspect of reality itself.

As we continue to explore the depths of the physical world, the universal constants that govern its behaviour will undoubtedly remain at the forefront of scientific inquiry. For in these enigmatic numbers may lie the keys to unlocking the most profound mysteries of our universe, from the origins of space and time to the nature of consciousness itself.

Physical constants are numerical values that define the laws of nature in the universe. Unlike variables, which can change depending on the situation or environment, constants remain fixed across time and space under given conditions. They represent the immutable laws that govern the behaviour of matter, energy, and forces.

Some of the most important physical constants include:

The Speed of Light (c): 299,792,458 metres per second. The speed of light is a universal constant that determines how fast light and all other electromagnetic waves travel through a vacuum. It plays a crucial role in the theory of relativity and the structure of spacetime.

Gravitational Constant (G): 6.67430×10^{-11} m^3 kg^{-1} s^{-2}. The gravitational constant determines the strength of the gravitational force between two masses and is a cornerstone of Newton's law of gravitation.

Planck's Constant (h): $6.62607015 \times 10^{-34}$ J s. Planck's constant governs the quantisation of energy in quantum mechanics, describing the relationship between the energy of a photon and its frequency.

Boltzmann Constant (k_B): 1.380649×10^{-23} J K^{-1}. The Boltzmann constant relates the average kinetic energy of particles in a gas to the temperature of the gas.

Elementary Charge (e): $1.602176634 \times 10^{-19}$ coulombs. This is the fundamental unit of electric charge, which quantifies the amount of charge carried by a single proton or the magnitude of the charge carried by a single electron.

Avogadro's Number (NA): $6.02214076 \times 10^{23}$ mol^{-1}. Avogadro's number defines the number of atoms or molecules in one mole of a substance.

Fine-Structure Constant (α): Approximately 1/137.035999 (dimensionless). The fine-structure constant is a dimensionless number that characterises the strength of the electromagnetic interaction between elementary charged particles.

These constants are the foundation upon which the laws of physics rest, ensuring that the same rules apply everywhere in the universe, regardless of conditions. But why do these constants exist? What is their origin, and why do they take the specific values they do? The existence of physical constants suggests that the universe operates according to specific, unchanging laws. These constants determine how particles interact, how forces behave, and how matter and energy evolve over time. Without them, the universe as we know it would be radically different—if it could exist at all.

For example, if the speed of light were different, the structure of space-time would be altered, and the principles of relativity would no longer hold. If Planck's constant were different, the energy levels of electrons in atoms would shift, changing the nature of chemical bonds and the behaviour of matter. Small variations in the gravitational constant would affect the formation of stars, galaxies, and planets, possibly preventing life from emerging in the universe.

Taken together, such sensitivities motivate what is often called the fine-tuning question.

Physicists often describe the fine-tuning puzzle in terms of **dimensionless ratios**—pure numbers that set the relative strengths of forces and the relative scales of masses—along with a few cosmological parameters. The basic intuition is simple: change the numbers and you change what the universe can *make*—whether it can still grow galaxies, sustain stars for long stretches, forge the heavier elements, and support the kind of stable, flexible chemistry life seems to need. Take dark energy: if the vacuum energy were much larger and positive, the universe could expand so insistently that matter would struggle to gather into galaxies; if sufficiently negative, it could collapse again before complexity has time to emerge. Take gravity: its relative weakness is part of what allows stars to shine steadily for billions of years. If gravity were much stronger relative to the other forces, stars would tend to form denser, burn through their fuel faster, and die younger—often ending sooner as compact remnants like white dwarfs, neutron stars, or black holes—leaving far less time for planets to form and mature. Take the strong nuclear force: change its strength and you change the key rungs on the fusion ladder—how readily hydrogen turns into helium, and how efficiently the later steps that build elements like carbon and oxygen can proceed. Even small shifts can make stars burn very differently and produce a different mix of elements. Take electromagnetism—captured by the fine-structure constant—and you begin to alter the size of atoms and the strength of bonds, nudging chemistry into a different world. And take carbon: the pathways by which stars make it depend on delicate details of nuclear structure, so shifts in those details can, in principle, change how abundant life's favourite element becomes. Put these together and you see why the question persists. The universe isn't just a container for matter; it's a workshop in which matter can organise into structure, chemistry, memory, and metabolism. What should we make of that? Some think a deeper theory will ultimately pin these numbers down. Others point out a quieter fact: we could only be here to notice the values in a universe where beings like us can arise. Either way, the takeaway is hard to shake—these constants do not merely describe reality; they help decide what kinds of reality are possible. What, if anything, should we conclude from this?

The existence of physical constants may point to the idea that the existence and constancy of these values may point to deeper informational principles, though this remains an open question. These constants represent fixed values that determine how information flows and how physical processes unfold. They are the rules by which the universe operates, ensuring that information is processed in a consistent, predictable manner. In a sense, physical constants can be seen as the parameters of a vast, cosmic computational system, where information is the fundamental currency. Just as the rules of a computer program determine how data is processed and what outputs are generated, the physical constants of the universe determine how particles interact, how forces behave, and how the universe evolves over time.

Constraints don't just limit; they channel—and life exploits those channels.

5.6 Entropy, Life, and the Information–Order Paradox

This perspective is consistent with the philosophical proposal that information may function as a kind of underlying principle or proto-force—an idea far from scientific consensus but gaining traction as a conceptual framework in some areas of theoretical physics. If the universe is indeed a form of cosmic computation, then the physical constants represent the underlying code that governs the entire system.

The existence of constants in the physical world is one of the most profound and intriguing aspects of nature. These constants—such as the speed of light, Planck's constant, and the gravitational constant—form the bedrock of our understanding of the universe and its laws. Their presence suggests that there is an underlying order or structure to reality, one that is governed by fundamental principles. The idea that these constants are fixed and unchanging across the universe invites deep philosophical questions about the nature of reality, and in particular, whether they point to a deeper informational or computational structure at the core of existence.

Physical constants provide a foundation of stability that allows the laws of physics to operate consistently across space and time. They define relationships between various quantities (e.g., mass, energy, charge) and ensure that these relationships hold universally. Without such constants, the behaviour of physical systems would be unpredictable, and the structure of matter, energy, and space-time could vary in different regions of the universe, making the coherent evolution of systems—such as stars, planets, and life—impossible.

This consistency is remarkable and suggests that the universe operates according to a set of underlying rules or algorithms. The existence of these constants gives rise to the idea that the universe might be governed by something akin to a "cosmic code" or "programme," where the constants act as parameters within this overarching system. These rules not only describe physical processes but also constrain them, leading to the predictable and stable behaviour of matter and energy across the cosmos.

The idea that these constants might reflect a deeper informational fabric of reality is increasingly being explored in both physics and philosophy. In modern physics, particularly in theories like quantum mechanics and information theory, there is a growing recognition that information may be as fundamental to the universe as matter and energy.

Here are a few ways that physical constants could point to a larger framework of information:

Quantum Information and Planck's Constant: In quantum mechanics, Planck's constant h is central to understanding the quantisation of energy, defining the smallest possible unit of action. Quantum mechanics itself can be seen as a theory of information—how information about a system is stored, transferred, and manipulated at the smallest scales of reality. The fact that h is a constant implies that there is a minimal "resolution" to information in the universe, suggesting that the fabric of reality may be "pixelated" at some fundamental level.

Speed of Light and Relativity: The speed of light c, another constant, defines the maximum rate at which information can travel through space. This sets an absolute limit on the transfer of information across the universe. In this sense, c is not just a speed limit for physical objects but a boundary on how quickly information about physical systems can propagate. It reflects the relationship between space, time, and information flow.

Entropy and the Second Law of Thermodynamics: The second law of thermodynamics, which governs the increase of entropy (a measure of disorder or information uncertainty) in isolated systems, is closely related to constants like Boltzmann's constant $k_{B.}$ The growth of entropy can be seen as a process of losing or dispersing information about a system's microstates. This, too, points to a fundamental relationship between information and the laws of physics.

Gravitational Constant and the Structure of Space-Time: Einstein's theory of general relativity, which describes how matter and energy curve space-time, is built upon the gravitational constant G. The structure of space-time itself can be viewed as encoding information about the distribution of mass and energy in the universe. The constancy of G ensures that this structure behaves in a uniform way across the cosmos, suggesting that space-time itself may be part of a larger informational framework.

The constancy of these values across the observable universe raises the possibility that they are not arbitrary but are instead a reflection of a deeper, more fundamental layer of reality. This idea has been explored in various ways:

a. The Simulation Hypothesis: Some philosophers and scientists have speculated about the "simulation hypothesis"—the idea that our universe might function like a vast computation, with physical constants akin to algorithmic parameters. While intriguing, this remains speculative and should not be conflated with the holographic principle, which has a separate theoretical basis in physics. In this view, reality itself could be a kind of digital construct, with the constants setting the rules for how information is processed in this simulation. The apparent regularity and precision of these constants lend some credence to the idea that they might be part of a larger "programme" that governs the unfolding of physical phenomena.

b. The Holographic Principle: In certain areas of theoretical physics, such as string theory and black-hole thermodynamics, there is a growing belief that the universe might be holographic in nature. According to this view, the information contained in a given region of space can be fully described by the information encoded on its boundary, much like a hologram. If the universe itself is a holographic projection of information, then the physical constants could be constraints on how this information is encoded and decoded across different scales of space and time.
c. Spiritual and Philosophical Interpretations: Many spiritual and philosophical traditions have long speculated about the existence of an underlying order or "code" to the universe. The idea of cosmic constants governing reality resonates with the concept of a divine or universal intelligence that has set the parameters for the universe to function. For example, in some interpretations of Eastern philosophies like Taoism or Hinduism, the constants of nature could be viewed as expressions of a cosmic order or "Tao" that permeates all things. Similarly, in Western philosophy, thinkers like Plato proposed that there are eternal forms or patterns that underpin the material world, of which physical constants might be a manifestation.

The presence of multiple constants, each governing distinct forces and phenomena, has led physicists to search for a unified theory that might explain why these constants take the specific values they do. Efforts like string theory and the quest for a "Theory of Everything" aim to reduce the free parameters in our models, potentially uncovering the origins of these constants and revealing a more fundamental framework.

Understanding how these constants relate could be pivotal in achieving a unified theory, marking a major leap in physics. Each constant—Planck's h governing quantum resolution, the speed of light c limiting information transfer, Boltzmann's k relating to information dispersal, and the gravitational constant G shaping space-time—uniquely frames the universe's information processing. Uncovering their interconnections may reveal an underlying "informational fabric," binding the fundamental forces into a cohesive framework and drawing us closer to a unified theory of everything. Such a theory might suggest that these constants are not truly fundamental but emerge from deeper principles or symmetries, implying they are facets of a larger, all-encompassing structure of information that underpins not only the physical universe but also the essence of reality itself.

In essence, the existence of physical constants points to a stable, universal framework governed by deep, consistent rules. The stability and universality of these constants suggest an underlying fabric that may be informational in nature. Across quantum mechanics, relativity, and cosmology, these constants appear to define the parameters for how information is stored, processed, and transferred. This perspective—that constants reflect a larger, perhaps computational or informational reality—opens new pathways for exploration, intersecting science, philosophy, and even spiritual thought in the pursuit of understanding the cosmos.

The question of whether information might be considered a fifth force of nature is speculative yet profound, rooted in advanced developments in physics and information theory. Information plays a central role in our understanding of quantum mechanics, black-hole thermodynamics, and the structure of spacetime, governing the behaviour of particles, the flow of energy, and the universe's evolution. Physical constants, representing fixed laws, could be viewed as parameters within a cosmic informational system, ensuring the universe's consistent, predictable operation and pointing towards the deeper informational principles underlying the physical world.

While information has not yet been formally classified as a force in the same sense as gravity or electromagnetism, its influence on the behaviour of the universe is becoming increasingly apparent in scientific discourse. As our understanding of quantum information theory and the nature of reality progresses, it is conceivable that future research may reveal information to be a principle as fundamental as the four known forces—perhaps even functioning as a kind of organising substrate underlying the cosmos.

One of the most fundamental ways in which information ties into the natural world is through the concept of entropy. The second law of thermodynamics states that the entropy of a closed system tends to increase over time. Entropy, in this sense, is often interpreted as a measure of disorder, reflecting the tendency of systems to move from order to chaos. However, entropy can also, in information theory, be interpreted as a measure of missing information about the state of a system. In physics, these interpretations are closely related but not always identical. The higher the entropy, the less information is available to precisely describe the system's configuration.

In this light, entropy is a kind of "cosmic ignorance," a measure of how much information about a system is hidden from us. This perspective provides a fascinating view of the universe as a battleground between order and disorder, where information flows continuously. Systems such as stars, planets, and even life itself can be seen as organised entities that manage to reduce local entropy, creating "pockets of order" in a universe otherwise dominated by chaos.

By connecting entropy to uncertainty in his landmark work "A Mathematical Theory of Communication" (*Bell System Technical Journal*, 1948), Claude Shannon, the father of information theory, crafted a mathematical framework to explain the transmission, storage, and loss of information.

Shannon's theories, though designed for communication systems, have far-reaching implications for how we understand information in the context of physical systems. In fact, they suggest that information is deeply intertwined with the very fabric of the universe.

The idea that the universe evolves through increasing complexity in the organisation of information is supported by the history of cosmic evolution. From the Big Bang to the formation of galaxies, stars, and eventually life, each stage can be seen as a reorganisation of information into increasingly complex structures. Life, in particular, appears to resist entropy by creating highly organised systems that encompass and manage vast amounts of information, such as DNA, neural activity, or cultural artefacts.

This leads to a radical hypothesis: What if information, rather than being a mere byproduct of physical processes, is a driving force in the universe? Just as gravity shapes structure, could information act as a bias towards organised complexity—an arrow of admissible transformations—within the overall rise of entropy? But if entropy reigns and order is ever-fleeting, how then does the universe persist in sculpting galaxies, stars, and life itself from a sea of randomness? What mysterious principles allow pockets of exquisite complexity to bloom amidst the cosmic drift towards disorder? Here, we must descend into the deep logic of thermodynamics—a logic that not only circumscribes the boundaries of existence but, in its very constraints, enables the improbable miracle of life.

The second law of thermodynamics, which states that entropy in a closed system tends to increase over time, is often understood as a fundamental principle of nature. Entropy, in this sense, measures disorder, demonstrating how systems shift from order to chaos or equilibrium. As time progresses, entropy naturally increases, leading systems towards randomness and decay. This principle is why things in the world eventually break down, why heat dissipates, and why all physical processes tend towards equilibrium. A hot cup of coffee cools down, a living organism dies and decomposes, and complex structures erode over time.

5.7 The Paradox of Order in an Entropic Universe

Yet, despite the overarching principle of increasing entropy, the universe displays countless examples of structure, order, and complexity. Although the universe tends towards disorder, order can still emerge and persist in various phenomena, from the stable orbits of planetary systems to the intricate processes that sustain life. This raises profound questions: If entropy increases over time, what allows for the emergence of such order? What forces or principles resist the pull of entropy, and how do they work? To answer these questions, we must examine not only the universal law of increasing entropy but also the rare and precious circumstances under which order can persist, especially in the context of life. Recent research across cosmology, astrobiology, and molecular biology converges on a remarkable insight: the emergence and survival of complexity are not accidents, but outcomes of precise thermodynamic balances, written into the structure of our universe itself. These balances are not theoretical curiosities—they manifest in the geography of our own cosmic address. The conditions that allow information and complexity to persist are written in the stars, the location of our solar system, and even in the quiet stretches between galaxies. This chapter explores these questions, focusing on how information, energy, and natural forces lead to the emergence and persistence of order, particularly in biological systems, planetary formations, and other complex structures. To answer these questions, we must look not only to the abstractions of physics but also to the concrete details of our cosmic address. Recent research across astronomy, astrobiology, and thermodynamics reveals that the very *placement* of life—where it is able to arise and persist—is far from random. The universe, it turns out, offers only the narrowest of thermodynamic windows in which complexity, information, and life can sustain themselves against the relentless drift towards disorder. The emergence and persistence of life in the universe appear tightly constrained by the principles of thermodynamics, especially the relationship between energy flux, entropy, and the maintenance of ordered complexity. Analysis of our own galactic environment demonstrates that the spatial context of the Solar System is not arbitrary. As described by Lallement et al. (2003), the Solar System traverses the Orion Spur—a minor interarm region of the Milky Way—occupying a region of reduced stellar density and lower cosmic ray flux known as the Local Bubble. Our location within the Local Bubble and Orion Spur implies reduced nearby dense-cloud encounters and supernova rates compared with inner-disc environments, potentially moderating astrophysical hazards (Lallement et al. 2003; Frisch et al. 2011). Studies by Lineweaver, Fenner, and Gibson emphasise that proximity to the galactic centre, with its elevated supernova rate and gravitational disturbances, increases the risk of biological disruption through radiation-induced mutagenesis and atmospheric loss (Lineweaver, Fenner, and Gibson 2004). The solar system's relatively tranquil position is mirrored at even larger scales: the Milky Way itself is not situated within a dense galactic cluster or an active region of frequent interactions but rather is located in a relatively quiet, low-to-moderate density filament of the cosmic web, on the outskirts of the Virgo Supercluster and adjacent to a structure known as the Local Void, as described by Tully et al. (2014).

This pattern of spatial isolation—from solar system to galaxy—reflects a broader principle in cosmic evolution: environments characterised by moderate densities and reduced dynamical interactions are statistically more likely to sustain long-term stability. Such conditions limit the frequency of catastrophic events, including close stellar encounters and high-energy collisions, thereby providing a persistent backdrop against which complexity can accumulate.

Thermodynamic principles clarify this distribution. Boltzmann grounded entropy in state counting: entropy increases when a system has access to more

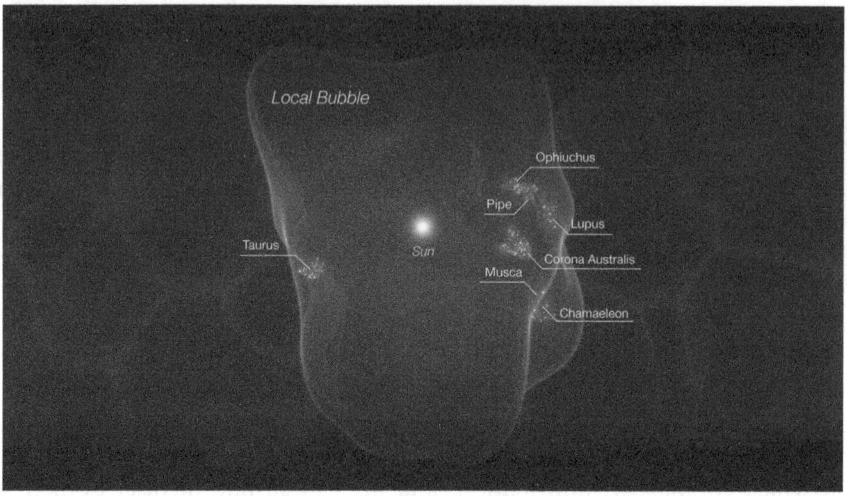

FIGURE 5.2 Sun's location inside the Local Bubble; nearby star-forming regions labelled.

microstates; high energy density alone does not fix entropy without specifying constraints and distributions (Boltzmann 1872; Callen 1985). In the context of astrobiology, this means that the intense radiation and gravitational perturbations in the inner Milky Way likely drive local systems towards maximal entropy, precluding the persistence of low-entropy, information-rich structures required for biochemistry. Nonequilibrium thermodynamics shows that open systems can sustain low-entropy structures by exporting entropy to the environment; what matters is the match between fluxes and dissipation pathways (Prigogine 1967; Nicolis and Prigogine 1977).

While there is no direct theory in physics stating that greater mass per se increases entropy independently of energy, it is established that mass concentrations—such as in stars, black holes, or dense planetary cores—result in extreme thermodynamic environments. For Schwarzschild black holes, entropy is proportional to horizon area $S \propto M^2$; by contrast, planetary habitability depends on multiple geophysical parameters beyond mass alone (Bekenstein 1973; Zahnle and Catling 2017). Life, as presently known, only arises on planets of moderate mass and temperature, where energy input from the host star is neither overwhelming nor insufficient.

At the biochemical level, living systems are information-dense: sequence-encoded polymers (DNA, RNA, proteins) and structured membranes store and process signals, but only within a narrowly controlled chemical milieu. The canonical elements—carbon, hydrogen, oxygen, nitrogen, phosphorus, and sulphur—are not "trace" so much as tightly buffered in specific stoichiometries and concentrations, with dissolved ions (e.g., H^+, Na^+, K^+, Ca^+, Cl^-) held within narrow physiological ranges. As laid out in *Molecular Biology of the Cell*—see "Protein Structure and Function" (solvent conditions and folding), "Membrane Transport" ($\frac{Na^+}{K^+}$ gradients and homeostasis), and "Electrical Properties of Membranes" (resting and action potentials)—cellular physiology depends on the precise regulation of ionic strength and pH; departures in either direction disrupt protein stability, membrane potentials, and genome maintenance (Alberts et al. 2015). From a thermodynamic perspective, this stringent homeostasis expresses the broader point that low-entropy, information-rich states are fragile and require continuous energy flow to resist degradation (Schrödinger 1944).

The distribution of life within planetary and aquatic systems further illustrates this principle. Aquatic species display higher reproductive success and population density in regions of low turbulence and moderated nutrient inflow, such as riverbanks and protected coastal zones, as shown by studies in freshwater ecology (Allan and Castillo 2007). Similarly, thermophiles at hydrothermal vents occupy niches defined by steep thermal and chemical gradients, with fitness peaks at intermediate distances from fluid sources (Jannasch and Mottl 1985; Baross and Hoffman 1985).

Extending this view beyond the Solar System, the Milky Way resides in a relatively low-density environment rather than within the crowded centres of galaxy clusters. This placement likely reduces disruptive gravitational interactions and high-radiation episodes, thereby offering longer, more stable intervals for biological evolution (Peebles 1993; Tully et al. 2014). While this remains an extrapolation, the available cosmological and biological evidence together imply that both the origins and sustainability of life are promoted by stable, moderate, and low-entropy environments.

The synthesis across disciplines suggests that life is best characterised not simply as organised matter, but as an extremely high density of biological information, encoded and preserved under conditions of optimal—not maximal—energy and mass flux. Life represents a local reduction in entropy, made possible only where gradients and concentrations are finely tuned. Too much energy, too much mass, or too great a fluctuation, and the system reverts to disorder, losing its informational content.

From the relative voids that shelter our Solar System, through the Goldilocks conditions of habitable planets, down to micromolar ionic balances within cells, the rise and persistence of life are governed by thermodynamic constraints. Freshwater studies show that biomass and diversity peak under stable, intermediate-disturbance regimes that balance resource flux with perturbation (Allan and Castillo 2007; Connell 1978), consistent with non-equilibrium thermodynamics, in which open systems maintain order only when energy gradients are neither too weak nor too strong (Prigogine 1967). Taken together, these strands indicate that the distribution of life reflects deep physical constraints linking mass, energy, and information rather than mere chance or chemistry—a connection anticipated by Schrödinger (1944) and elaborated in modern cell biology (Alberts et al. 2015). In this view, life is locally maintained, information-dense order that persists only within narrow thermodynamic windows. As Peebles (1993) details in *Principles of Physical Cosmology*, the same logic that governs the distribution of galaxies and the cooling of stars also scripts the story of cellular order and biological resilience, a concept linked to energy dissipation and order maintenance in Schrödinger (1944). As emphasised by Prigogine (1967), thermodynamics does not merely delimit the possible—it shapes the actual, writing the fragile script by which life maintains itself against the gathering tide of chaos. In every atom and every star, this balance is continuously renegotiated. Order is not a static achievement but an ongoing performance—one that demands constant rehearsal, innovation, and adaptation. In the next act, order itself becomes not just a product of the universe but its creative engine.

And yet, the story does not end with the emergence of order. For once born, order itself becomes an engine—processing information, organising matter, and creating ever more intricate patterns, as suggested in the information-theoretic approaches to complexity by Lloyd (2006). From the clustering of galaxies to the coiling of DNA, the interplay between entropy and information continues, producing systems that not only resist disorder but also generate new forms of complexity. This cosmic perspective reframes our understanding: order, complexity, and life do not arise in defiance of thermodynamics but flourish precisely where physical laws—mass, energy, entropy—are balanced with exquisite precision. At every scale, from the galactic to the molecular, it is information—encoded in structure, maintained by gradients, preserved by stability—that enables life to carve out islands of order in a sea of entropy.

The second law of thermodynamics suggests that isolated systems naturally tend towards increasing disorder or entropy. Given this principle, one might expect the universe, with its countless masses interacting through gravity and other forces, to eventually break down into chaos. Surprisingly, however, we observe remarkable order, structure, and stability throughout the cosmos. This raises a question: How can order arise in a universe governed by entropy?

One of the main forces responsible for organising matter is gravity. While gravity might intuitively seem to encourage a chaotic collapse of matter, it instead creates stable structures such as stars, planets, and galaxies. These systems are not static; they remain dynamic, held together in stable orbits and configurations. Planetary systems, for instance, exhibit a long-term stability in which planets orbit stars, and moons orbit planets in a well-defined, orderly manner that has persisted for billions of years. This intricate balance arises as gravitational forces pull matter together, creating structured systems that release energy in the process—resulting in pockets of localised order.

Life represents an extraordinary example of order in a universe that trends towards disorder. Living organisms maintain complex organisation at multiple levels—molecular, cellular, and systemic—despite the natural push towards entropy. A single cell, for example, is a highly organised structure composed of macromolecules that interact in precise ways to sustain life. When an organism dies, its structure breaks down, increasing entropy. However, as long as life persists, organisms counteract this tendency by constantly exchanging energy with their surroundings. They consume food, water, and air, using this energy to grow, repair, and reproduce. This continuous intake and use of energy allows life to defy entropy by maintaining order over time.

Biological growth and development further highlight how complex order can emerge from simple beginnings. An embryo, guided by the genetic information in its cells, organises into a fully developed organism with highly ordered tissues and organs. DNA encodes the instructions that drive growth, cellular differentiation, and the formation of functional structures. Despite the universal trend towards disorder, genetic information organises cellular activities, creating complexity from simplicity.

The formation of galaxies, stars, and planets also exemplifies this dynamic of order arising within a universe subject to entropy. Gravitational interactions between particles and masses lead to the formation of stars, which convert hydrogen into helium through fusion, radiating vast amounts of energy. Planets emerge and fall into stable orbits around stars, creating solar systems that can persist for billions of years. These localised regions of order demonstrate that, although the universe tends towards overall entropy, pockets of structure and stability emerge due to the fundamental forces at work.

As articulated by Shannon (1948) and by Schrödinger (1944), underlying these examples is the concept of information, which serves as a blueprint that organises matter and energy into structured systems. In living organisms, DNA encodes the instructions necessary for growth, repair, and metabolism, allowing life to counteract entropy by harnessing and directing energy flows. In physical systems, patterns of information can determine the arrangement of molecules, atoms, and larger structures like galaxies. This organisation results not only from natural forces but also from information that shapes the behaviour of matter and energy, guiding them into forms that persist over time.

Beyond biological systems, information also drives order in non-living systems. Snowflakes, for example, form through the orderly arrangement of water molecules into symmetrical structures. Crystals, too, exhibit highly organised atomic patterns based on the specific molecular properties of the material. Even in chemical reactions and fluid dynamics, complex patterns of self-organisation emerge. These systems still adhere to the second law of thermodynamics, yet they demonstrate how information and energy exchange can create temporary pockets of order even within a framework that trends towards disorder.

The concept of open systems helps explain how order can emerge in a universe governed by entropy, as first formalised by Prigogine (1967). While closed systems inevitably move towards disorder, open systems—those that exchange energy and matter with their surroundings—can sustain or even increase their level of order. Living organisms, planetary systems, and galaxies are examples of open systems, continually interacting with their environment. For instance, plants harness energy from sunlight to drive photosynthesis, animals consume plants to fuel their own processes, and even stars release energy into the cosmos. This constant energy flow enables these systems to maintain order and resist entropy's pull.

The interplay of information, energy, and natural forces is essential in shaping the universe's organised structures. Information acts as an organising principle, while energy sustains order, and fundamental forces like gravity and electromagnetism guide matter's behaviour. This balance creates temporary yet resilient structures that counter entropy, leading to the emergence of complex systems.

As life evolved, it developed increasingly sophisticated ways of processing information. The genome of an organism, for example, encodes information necessary for constructing and maintaining life, a relationship first quantitatively described in Avery, MacLeod, and McCarty's "Studies on the Chemical Nature of the Substance Inducing Transformation of Pneumococcal Types" (*Journal of Experimental Medicine*, 1944). Evolution optimises information processing—genes that are better at encoding, replicating, and utilising information have a higher likelihood of survival. This trend towards refined information processing is evident in structures like the eye, which evolved to capture and interpret visual information with remarkable efficiency.

At a fundamental level, life can be viewed as an ongoing effort to counter entropy by processing information and energy to maintain order. This process requires a continuous intake of information, whether in the form of sensory inputs, cellular signalling, or genetic replication. Life is thus not merely about survival; it's about sustaining order through efficient information processing.

Expanding this concept to the universe itself, theoretical physicist John Archibald Wheeler proposed the idea of "It from Bit," suggesting that every physical entity originates from units of information—bits (Wheeler 1989). In Wheeler's view, information is the building block of reality, giving rise to physical structures such as atoms, stars, and biological systems. This notion represents a shift in perspective, where matter and energy are not the most fundamental components but are instead emergent properties of informational interactions.

Quantum mechanics offers compelling evidence for the role of information in shaping what outcomes are possible and how probabilities are assigned. At the quantum level, particles exist in multiple states until observed, suggesting that the act of measurement—or processing information—determines the outcome. Quantum entanglement further illustrates this, with measurement outcomes showing nonclassical correlations across distance—stronger than any local-realist theory allows, yet without enabling faster-than-light signalling. This phenomenon hints that information transcends conventional space and time, challenging traditional notions of causality.

If information is indeed foundational, the laws of nature could be seen as rules that govern information processing throughout the universe. Constants like the gravitational constant and the speed of light might represent aspects of the universe's informational framework, organising how mass, energy, and space interact. In this view, the physical universe operates as a vast computational system, with the laws of physics functioning as its programming code.

This computational interpretation suggests that the universe is not a static collection of particles and forces, but a dynamic, self-organising system driven by information flow and transformation. This perspective not only deepens our understanding of physics but also opens new avenues for exploring the fundamental nature of reality itself.

The idea that information is the fundamental substance of the universe touches on deep philosophical questions. Since ancient times, philosophers have grappled with the nature of reality, exploring whether the material world is truly the ultimate source of existence or whether deeper, more abstract principles govern the cosmos.

i. Plato's Theory of Forms and Information
Plato's theory of forms suggests that the material world is not the true essence of reality but rather a reflection of abstract, perfect forms or ideas. In the context of information theory, Plato's forms can be understood as the underlying informational structures that give rise to physical objects. Just as a chair in the physical world is an imperfect representation of the ideal "chair" in the world of forms, a physical object might be seen as an emergent property of the underlying informational blueprint that defines its structure.

ii. Pythagoras, another ancient philosopher, believed that reality is fundamentally mathematical, organised by numbers and ratios. This view aligns with the modern understanding of information as the foundation of reality. Just as mathematics underpins the structure of the physical world, information underpins the structure of reality itself.

iii. Advaita Vedanta and the Illusion of the Material World
Eastern philosophy also offers insights that resonate with the information paradigm. In Advaita Vedanta, an ancient Indian philosophical tradition, the universe is seen as a manifestation of Brahman, the ultimate, unchanging reality behind all phenomena. The material world is considered an illusion (Maya), and what truly exists is an underlying oneness—a concept that can be compared to the informational fabric of the universe.

In this view, the physical world, with all its diversity and complexity, is a manifestation of a deeper, unified reality. Viewing information as the fabric of the universe suggests it as the origin of all phenomena, with physical objects merely reflecting temporary configurations of this fundamental informational base.

iv. Teilhard de Chardin and the Omega Point
In the 20th century, philosopher and Jesuit priest Teilhard de Chardin proposed the idea of the Omega Point, a future state in which the universe reaches its maximum complexity and consciousness. For de Chardin, evolution is not just a biological process but a cosmic one, leading towards greater informational organisation and self-awareness.

De Chardin's vision of the universe evolving towards a higher state of consciousness aligns with the idea that information drives the evolution of complexity. Just as life on Earth has evolved to process information more efficiently, the universe as a whole may be evolving towards a state of maximum informational complexity, where consciousness plays a central role.

As we move further into the Information Age, humanity's relationship with information is changing in profound ways. The rise of computing, artificial intelligence (AI), and quantum computing represents a dramatic expansion of our capacity to process, store, and transmit information. In many ways, we are externalising our cognitive processes into machines, creating new forms of intelligence that extend beyond biological limitations.

AI has already begun to reshape industries and societies, demonstrating the power of machines to process information at speeds and scales far beyond human capability. As AI systems become more sophisticated, they may develop forms of consciousness or self-awareness, raising profound ethical and philosophical questions about the nature of intelligence and the role of humans in the cosmic information process.

With its ability to process vast amounts of information simultaneously through superposition and entanglement, quantum computing has the potential to revolutionise our understanding of both computation and the universe itself. Quantum computers could solve problems that are currently beyond the reach of classical computers, potentially unlocking new insights into the nature of reality and information.

As technology advances, the boundaries between human consciousness and machines may begin to blur. Brain–computer interfaces and mind-uploading technologies could allow humans to merge with the informational fabric of the universe, transcending biological limitations and accessing new forms of consciousness.

This raises profound questions about the future of humanity and the role we will play in the cosmic information process. Whether we will become agents in the universe's ongoing quest to understand itself or whether we are merely byproducts of a larger, more mysterious system of information flow.

Returning to the central question of this chapter: Could information be considered the fifth force of nature? As we have seen, information plays a fundamental role in shaping the structure and behaviour of physical systems, from the quantum level to the scale of galaxies. Just as gravity organises matter on a cosmic scale, information may organise energy and matter at all scales, guiding the evolution of the universe.

The quest for a theory of everything—a single framework that unites quantum mechanics and general relativity—has long eluded physicists. However, the idea that information is the fundamental building block of reality offers a promising avenue for unification. If information underpins both quantum particles and the curvature of spacetime, then a deeper understanding of information theory could reveal the connections between these seemingly disparate domains.

In this view, the constants of nature, such as the speed of light and the gravitational constant, might be seen as fixed points in an underlying informational structure. The behaviour of matter and energy, from the motion of planets to the interactions of subatomic particles, could be understood as manifestations of the rules governing how information is processed in the universe.

Information transcends mere communication or computation; it is the fundamental fabric of reality, the cosmic substrate from which all physical phenomena emerge and organise. From quantum bits and DNA to human language, information unlocks the mysteries of not only natural laws but also the essence of consciousness and existence itself. As we continue to explore the role of information in physics, biology, and philosophy, we may come to see information as the missing link that unifies our understanding of the universe and its underlying principles. It seems that information is not merely a descriptor of reality but also a driving force that shapes the cosmos, propelling the evolution of complexity, consciousness, and life.

In this sense, we are not separate from the universe but rather molecules observing ourselves through the lens of information processing. This insight may be the key to unlocking the deepest mysteries of life, consciousness, and reality itself. Information, pervasive and structuring, acts as a guiding principle that shapes the cosmos—a potential "fifth force" in the sense of an organising substrate that may help answer the fundamental questions of existence. Yet, even as information weaves the visible tapestry of order, a profound tension remains. The universe, governed by laws both rigid and enigmatic, seems intent on unravelling its own creations. How does information—delicate, structured, and improbably enduring—persist in a cosmos so relentlessly tilted towards disorder? To answer this, we must descend into the deep logic of thermodynamics: a logic not of static rules, but of continual negotiation between chaos and coherence.

The concept of information as a central force is not only profound but also supported by leading scientific thought. Leonard Susskind, a pioneer in theoretical physics, has contributed significantly to our understanding of this idea, especially through his work on the holographic principle and black-hole complementarity, which are central to modern debates about black-hole information. One of his most influential assertions is the conjecture that information is never truly lost, even in black holes—a principle he defends through black-hole complementarity and the holographic principle. While this view is widely discussed, the information paradox remains a topic of active debate. This principle, which aligns with the conservation laws in physics, reinforces the notion that information is woven into the very structure of the universe, resonating with the foundational principles of quantum mechanics, general relativity, and thermodynamics.

By embracing information as an intrinsic aspect of reality, we can begin to see the universe not as a collection of isolated phenomena but as an interconnected web where every particle, field, and force operates in harmony, guided by an underlying informational framework. This perspective opens new avenues in fields from quantum computing to neurobiology, suggesting that understanding information's role could be our gateway to comprehending consciousness, evolution, and perhaps even the nature of existence itself.

5.8 The Black-Hole Information Paradox: Susskind, Holography, and the Conservation of Information

In what follows, we begin with the black-hole information paradox and Hawking's discovery of radiation, turn to Susskind's response—black-hole complementarity and the holographic principle—then step back to the role of information in quantum mechanics, the thermodynamic arrow of time, and the intersection of entropy and information in black holes. We then widen the lens to ask whether information plays a foundational role across physics and close with the broader implications of that view.

First, we consider the black-hole information paradox, tracing Hawking's discovery of thermal radiation and the apparent tension it creates with unitary quantum dynamics and information conservation. Black holes, regions of spacetime with gravitational forces so intense that nothing—not even light—can escape, were first predicted by Einstein's theory of general relativity. As fascinating as these objects are, they pose a significant problem when examined through the lens of quantum mechanics, specifically with respect to information.

To make that tension precise, recall the quantum requirement of unitary evolution and information conservation. The paradox arises from the fact that, according to quantum mechanics, information about a physical system must always be preserved. This is rooted in the unitary evolution of quantum systems, where information is encoded in quantum states that evolve predictably. However, when something falls into a black hole, all observable information about the object is seemingly lost once it crosses the event horizon (the boundary beyond which nothing can return). If the black hole eventually evaporates completely via Hawking radiation, it seems as if the information is permanently erased from the universe.

Historically, this clash was sharpened by Hawking's argument that black holes could erase information. This contradiction between the principles of quantum mechanics and the classical description of black holes is known as the black-hole information paradox, first formally articulated by Hawking (1976). It has been one of the central puzzles in theoretical physics for decades, with many physicists, including Hawking himself, initially arguing that black holes could indeed destroy information—a claim that would violate the fundamental tenets of quantum theory.

The roots of the paradox trace back to Hawking's original calculation of black-hole radiation. In 1974, Stephen Hawking proposed that black holes are not completely black but emit thermal radiation due to quantum effects near the event horizon, a phenomenon now known as Hawking radiation. According to his theory, particle–antiparticle pairs are constantly being created near the event horizon of the black hole. Occasionally, one particle falls into the black hole while the other escapes, causing the black hole to lose mass over time and eventually evaporate completely.

Precisely because that radiation is thermal, however, the puzzle deepens. While Hawking's discovery was revolutionary, it deepened the information paradox. The radiation emitted by a black hole is purely thermal, meaning it contains no information about the material that originally fell into the black hole. This suggests that when the black hole eventually evaporates, the information about everything it swallowed is lost, leading to a violation of quantum mechanics' principle of information conservation.

A leading response reframes where the information resides rather than conceding its destruction. Leonard Susskind, together with Gerard 't Hooft and others, aimed to solve the black-hole information paradox. One of his key contributions was the proposal of black-hole complementarity, introduced by Susskind, Thorlacius, and Uglum (1993), which posits that information about objects falling into a black hole is not lost but instead is encoded in two distinct ways: one for an outside observer and another for an observer falling into the black hole. This avoids the paradox by suggesting that information is not destroyed, but different observers perceive it in different ways.

This viewpoint dovetails with the broader holographic idea that bulk information is encoded on boundaries. Another crucial idea championed by Susskind is the holographic principle, first introduced by 't Hooft (1993) and further developed by Susskind (1995). This principle suggests that the information contained within a volume of space can be represented as encoded on the boundary of that space, similar to how a 3D object can be represented as a 2D hologram. In the case of black holes, the holographic principle implies that all the information that falls into a black hole is encoded on its event horizon, a two-dimensional surface.

Taken together, these insights suggest that information is preserved, albeit highly scrambled, at the horizon. This radical idea fundamentally changed our understanding of information and black holes. It suggests that the seemingly paradoxical loss of information is not real—information is preserved, but it is stored on the event horizon in a highly scrambled form, which might eventually be recovered.

Crucially, this resolution respects quantum theory's core commitment to unitary dynamics. Susskind's resolution to the black-hole information paradox aligns with the deep principles of quantum mechanics. In quantum theory, information is stored in the state of a quantum system and can never be lost due to the unitary nature of quantum evolution. Quantum entanglement and the superposition of states illustrate how information is preserved, even when it becomes highly distributed or "hidden" from certain perspectives.

A familiar illustration comes from entanglement, where information persists even when it is widely distributed. For example, in quantum entanglement, two particles share a quantum state such that measuring one particle instantaneously provides information about the other, regardless of distance. Even if one of the particles is destroyed or absorbed, the information about the original system persists in the remaining particle. This principle supports

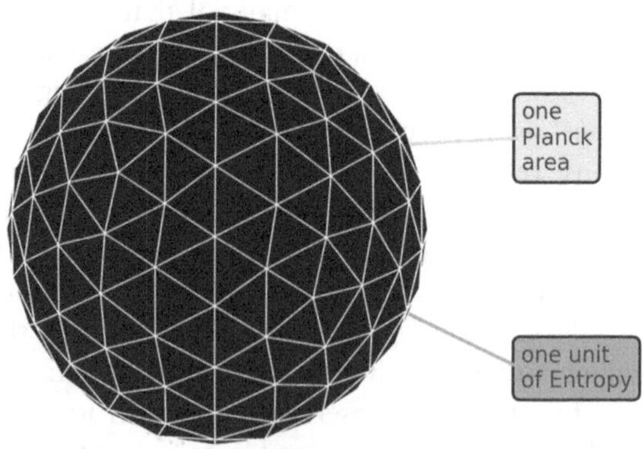

FIGURE 5.3 Black-hole event horizon as an information screen. The triangulated sphere sketches Planck-scale "pixels" on the horizon. The yellow cell marks one Planck-area patch; the blue cell marks one unit of entropy/bit stored on the surface. Entropy scales with area, not volume.

Susskind's view that information may be hidden in black holes but is never destroyed.

Thermodynamics points in the same direction when read through black-hole entropy. Susskind's argument also has profound implications for our understanding of thermodynamics, particularly the second law, which states that entropy, or disorder, in a closed system must always increase. In the context of black holes, the entropy is proportional to the area of the event horizon, as discovered by Bekenstein (1973) and extended by Hawking (1974). This is known as Bekenstein–Hawking entropy.

On that reading, the horizon area counts the available microstates and thus the information capacity. If black holes are to obey the laws of thermodynamics, then their entropy must reflect the information they contain. Susskind's holographic principle supports this by suggesting that the information is stored in the event horizon, and thus the entropy represents not just the mass of the black hole but also the information about everything that has fallen into it.

Even so, subtleties remain about where and how that information is stored and recovered. The relationship between entropy, the arrow of time, and information in black holes is subtle. While entropy defines a statistical direction to time's flow, how information is stored or retrieved from black holes is still a subject of ongoing theoretical development.

Viewed more broadly, black-hole thermodynamics hints at a general truth about where information exists in spacetime. Bekenstein's proposal that black-hole entropy is proportional to the area of the event horizon, rather than its volume, was one of the first hints that black holes encode vast amounts of information on their surfaces. Susskind extended this idea through his holographic principle, arguing that black-hole thermodynamics demonstrates a more general truth about information in the universe: it is not contained in the volume of space but in its boundaries.

This shift has been transformative for how we think about space, time, and quantum gravity. This insight has been transformative, leading physicists to reconsider the nature of space, time, and information in the broader context of quantum gravity. It suggests that black holes are not exceptional objects but are instead deeply connected to the fundamental workings of the universe.

The upshot is that information functions as a core constituent of physical theory, not a mere bookkeeping device. Susskind's work underscores the idea that information is a fundamental component of the universe, as important as matter, energy, and the forces that govern them. Information is the bedrock of quantum mechanics, where the evolution of quantum states is governed by the laws of physics in such a way that information is never lost. This principle is also foundational in other areas of physics, including thermodynamics, where entropy represents the flow of information in a system.

The preservation of information in black holes, as Susskind argues, supports the idea that information is a fundamental aspect of reality. Whether it is encoded in the quantum states of particles or in the event horizons of black holes, information persists.

The implications of Susskind's ideas extend far beyond black holes. If information is never truly lost, even in the most extreme environments, then the universe itself can be viewed as an intricate web of information processing. This perspective has influenced a range of fields, from quantum computing to cosmology.

In quantum computing, the preservation and manipulation of information are central to developing new computational paradigms that harness the principles of quantum mechanics. In cosmology, the holographic principle has provided a new framework for understanding the structure of the universe, where the three-dimensional world we experience may be a projection of more fundamental two-dimensional information encoded on the cosmic horizon.

In this view, the universe may be usefully conceived as an enormous information processor, constantly encoding, transmitting, and transforming information—though whether this is literal or merely a fruitful metaphor is a subject of philosophical debate. Whether in black holes, particle interactions, or the expansion of space, information remains a fundamental and indestructible aspect of reality.

6

TIME AS THE SCULPTOR OF INFORMATION

6.1 Time as the Hidden Sculptor: Beyond Energy and Entropy

Across the long arc of existence, humanity has sought to understand what sculpts reality—what chisels structure from chaos and breathes coherence into the void. For centuries, the dominant candidate has been energy: the currency of work described by Boltzmann in *Vienna's Academies* in the 1870s, the engine of stars as revealed by Eddington in the 1920s, and the fuel of change animating the cosmos as Feynman so elegantly demonstrated in his lectures at Caltech in the 1960s. In biology, energy drives metabolism and growth; in *What Is Life?* Schrödinger famously suggested that living systems "feed on negative entropy," maintaining local order by exporting disorder to their surroundings (Schrödinger 1944). In industry, it powers machines and production. But what if there exists a deeper sculptor? One that works not with heat or force but with pattern, reduction, and efficiency—a sculptor whose true substrate is not substance, but time?

Before we proceed, let us clarify what is meant by time in this context. In physics, time is the measurable duration between events—the fourth dimension of spacetime (Minkowski 1908); in information theory, time appears as rate and latency constraints on encoding, transmission, storage, and learning—the cadence at which signals are sampled, processed, and decoded (Shannon 1948). Physical time is what the cosmos runs on; informational time is what thought and evolution dance to. This chapter will explore how these two senses of time, though distinct, intertwine in the sculpting of reality. Physical time sets the boundaries within which all systems must operate, but it is within those boundaries that informational time emerges—expressing itself most fully in systems capable of processing, refining, and compressing

DOI: 10.4324/9781003740766-6

information. In this sense, informational time is not separate from physical time, but, rather, it arises from the physical substrate as complexity increases.

This chapter proposes a radical reframing. While energy governs transformation, time governs refinement. And what time refines is not just matter—it is information. The true evolution of systems—biological, cultural, technological—is not a tale of accumulating energy but of compressing information over time, a principle hinted at by Wiener's early work on cybernetics at MIT in the 1940s. This process, repeated across epochs and domains, leads not only to survival but to elegance, as open, driven systems channel energy into persistent, low-entropy structures (Prigogine 1978). It gives rise to structure, signal, and meaning.

If entropy sets a statistical arrow for spreading and mixing, time supplies the arena for selection and refinement; open, driven systems can self-organise into long-lived, low-entropy structures that dissipate energy—an insight crystallised in the study of dissipative structures (Prigogine 1978). Over time, rivers straighten, languages shorten, societies adapt, and ideas crystallise. This is not merely a response to disorder. It is the imprint of a deeper law: information wants to flow more efficiently through time.

In this chapter, we move away from entropy as the main protagonist and instead place time at the centre. We will explore:

How systems evolve towards efficiency not because of energy limitations but due to time-bound constraints.

- Why humans and nature alike prefer shortcuts, simplifications, and faster feedback loops.
- How language, technology, and even memory reveal time's role in compressing complexity.
- And, ultimately, how time and information may not be separate variables at all but two dimensions of the same principle.

Time is not just what passes. It is what edits. It does not merely erode—it also refines. Just as a sculptor chips away marble to reveal form, time erases redundancy to expose essence. And in that revelation lies the story of everything that endures.

Having reframed time as the hidden editor behind all refinement, let us examine how its logic sometimes defies our intuitions—beginning with the counterintuitive paths of efficiency that shape both the cosmos and daily life.

6.2 The Paradox of Efficient Detours: When Longer Is Faster

Let us now begin that story—by pausing on a paradox at the very heart of nature's efficiency: the possibility that in a universe ruled by time, the shortest route to refinement is not always the straightest.

It is here, in the winding interplay between path and purpose, that the hidden logic of optimisation emerges most clearly.

In our intuitive experience of the world, speed is often synonymous with straightness; yet in curved spacetime the physically relevant paths are geodesics shaped by gravity, as formalised in general relativity and dramatically supported by eclipse observations (Einstein 1915; Dyson et al. 1920). The same geometry shows up in flight paths: in orbital mechanics, for instance, the straightest line is often not the quickest. The fastest journey between two places may require a curve, a swing, or a leap to a higher altitude. Welcome to the paradox of efficient detours.

One of the clearest examples of this is the Hohmann transfer orbit—a method used by spacecraft to move from one circular orbit to another using the least amount of fuel (Hohmann 1925). Rather than firing directly at a target in space, a spacecraft first moves to a higher elliptical orbit, coasting along a longer path that ultimately allows it to rendezvous more efficiently with its destination. This counterintuitive principle—that a longer route can consume less fuel (delta-v) even if it takes more time—reveals a fundamental truth: systems evolve to exploit the curvature of time and space, not just their distances.

Gravitational slingshots exploit a planet's motion to exchange momentum and boost spacecraft speed with minimal fuel—an approach analysed in the early 1960s and later used to enable multi-planet tours (Minovitch 1963). These techniques reflect nature's deeper principle: leverage time and structure to gain velocity. This is not just physics—it's a metaphor for all efficient systems.

Much like a planet's gravity bending a spacecraft's path for gain, human systems bend around constraints of time to optimise outcomes. This shows up not just in rocket science but in everyday life. Consider rivers.

A young river winds across the landscape in lazy, inefficient meanders. But over time, sediment deposits and erosion reshape its path, and it straightens. Why? Because water, too, seeks a faster route. A river is governed by time and gravity, chiselling its way to minimal resistance. Alluvial rivers typically *increase* sinuosity through meander growth until cutoff events create oxbow lakes and shorter, locally straighter reaches; over long times, channels evolve towards a dynamic equilibrium that balances slope, discharge, and sediment load—efficiency emerges statistically not as monotonic "straightening" (Leopold and Wolman 1957).

Footpaths in parks offer a humble but telling lesson. Designers pour concrete, and then pedestrians draw their own maps—cutting diagonal "desire paths" that trace not the official plan but the least-effort way to get from A to B. With repeated use, those shortcuts get worn into the ground, and planners often repave along them: the environment has been edited by experience. These trails are informational fossils—records of how a system (a city and

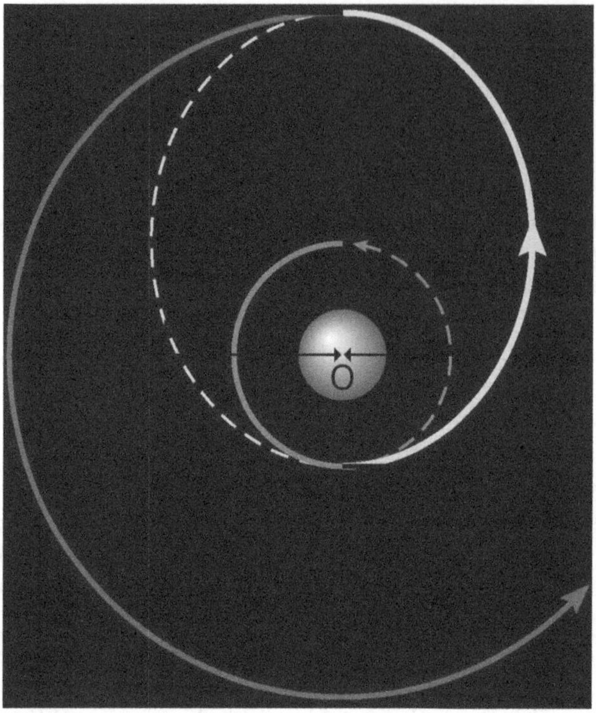

FIGURE 6.1 Hohmann transfer orbit: a two-impulse manoeuvre that moves a spacecraft from a lower circular orbit (1) to a higher circular orbit (3) via an elliptical transfer arc (2). A prograde Δv at departure injects the craft into the transfer ellipse; a second Δv at apogee circularises into the outer orbit. For coplanar, concentric orbits, this minimises total Δv (fuel).

its walkers) discovered and then committed to a better route under everyday time pressure. Far from being oddities, desire lines are typical of how rivers find channels, ants find food, packets find routes on the internet, and code paths get optimised in software: repeated use pushes motion towards efficiency.

6.3 Time as Coach and Editor: Temporal Pressure in System Optimisation

Time doesn't just wear systems down; it trains them. Given repetition and feedback, what starts crooked or inefficient—flows of water, crowds, computations—tends to reorganise into good-enough geodesics: paths that minimise relevant costs such as delay, energy, or risk. That idea is made precise in several formalisms. In dynamic programming, we compute policies that minimise cumulative future cost step by step (Bellman 1957). In optimal control, we pick control signals that extremise a performance functional subject to dynamics (Pontryagin et al. 1962). And in evolutionary dynamics, lineages that process energy and information more effectively are preferentially retained, while less efficient strategies fade (Turing's postwar work, though about pattern formation, shows how simple local rules can generate ordered spatial structure over time; Turing 1952).

The upshot is intuitive: temporal pressure tends to sculpt not the straightest line but the smartest curve. In physics, mass bends spacetime so that light follows a curved geodesic—dramatically confirmed during the 1919 eclipse (Dyson, Eddington, and Davidson 1920). In decision spaces, constraints and costs play the role of curvature, bending our routes through design, compression, and coordination. Under sustained use, systems learn the curve that saves what matters—time, energy, or error—and they etch it into the landscape.

In this light, we might redefine intelligence as the ability to take the most effective curved path through a complex terrain—spatial, cognitive, or conceptual. The smart agent, whether a neuron or a spacecraft, does not brute force through resistance. It flows with temporal efficiency—a principle evident in a principle consistent with Hebbian learning (Hebb 1949). And this tendency—visible from the trajectories of planets to the footfalls in a park—is the mark of information being sculpted by time.

Why do systems simplify? Why does nature carve winding rivers into straighter channels, or why do crowds invent new footpaths despite neatly arranged pavements? Why do languages evolve from elaborate syntax into snappier, more compressed forms? The answer, again and again, is time—a theme explored by Zipf's principle of least effort (Zipf 1949).

Time is not just a medium in which systems operate—it is a constraint against which they optimise. In a world where survival is measured in how swiftly and effectively one adapts, efficiency is not optional—it is evolution's favourite tool, grounded in natural selection (Darwin 1859). The longer a system exists within the constraints of time, the more likely it is to refine itself, trimming redundancies and streamlining its function. That is why older civilisations accumulate denser myths, older technologies compress more function into less space, and long-lived lineages display increasingly fine-tuned traits—yet biology remains as contingent as it is optimised. Gould (1989) reminds us that evolution is not a march to global perfection but a branching, historically path-dependent process. Consider neurological development in humans. A baby is born with an abundance of neural connections—more than it needs. Over time, these connections undergo a process called neural pruning, where rarely used synapses are eliminated and frequently used ones reinforced—a trajectory mapped in classic developmental studies (Huttenlocher 1979). This may sound like a loss, but it is optimisation. It is time to sculpt the brain to become more efficient, more responsive, more intelligent.

The same occurs in evolutionary biology. Structures that are not essential fade away. Fish living in lightless caves often lose functional eyes—a phenomenon synthesised in developmental and evolutionary work (Jeffery 2001). Birds in predator-free environments may lose flight, as Darwin noted in the Galápagos. Redundant genes can become pseudogenes, a dynamic process articulated in the theory of gene duplication (Ohno 1970). Time does not reward maximal complexity—it rewards optimal functionality. In nature, however, the paths carved are rarely mathematically optimal. Instead, evolution tends to find solutions that are "good enough" for survival—shaped by constraints, historical contingencies, and chance—rather than global perfection. And what we witness in the code of life, we find echoed in the codes we create. The drive for efficiency, once sculpted by nature, is now reenacted in silicon and software—an informational recursion spanning from neurons to nanometres.

In human-made systems, this same pressure is even more visible. The journey from room-sized vacuum-tube computers to microprocessors with billions of transistors etched into silicon wafers exemplifies this march towards computational efficiency. Moore's Law captured the empirical doubling of transistors per chip roughly every two years, while Dennard scaling explained how power density could remain constant as transistors shrank—until it broke down in the mid-2000s (Moore 1965; Dennard et al. 1974).

Here, time operates in both its physical sense—the calendar years through which technology marches—and its informational sense, as each new generation encodes more intelligence per computational step, as explored by Dennard et al. (1974). The history of computing is a record not just of physical progress but of an accelerating tempo in the language of information itself.

Today, we etch intricate circuits onto layered semiconducting materials—primarily silicon—encoding trillions of operations per second into a chip no larger than a fingernail. This is compression. This is efficiency carved by time. Language, too, bends towards brevity. Its evolution is shaped both by the physical constraints of speech—the seconds it takes to utter a thought—and by the informational limits of cognition, memory, and cultural transmission, as early structural linguistics noted (Sapir 1921). The pressure of time is felt in the muscles of the tongue as much as in the architecture of syntax. Sanskrit's highly structured grammar (Pāṇini) coexisted with evolving Prakrits and later Indo-Aryan vernaculars; likewise, Latin diversified into the Romance languages. Importantly, languages do not uniformly "simplify" over time—they redistribute complexity across morphology, syntax, and pragmatics depending on sociolinguistic pressures (Emeneau 1956; McWhorter 2011; Trudgill 2011). Latin, once the tongue of empires, gave rise to French, Spanish, and Italian—each more accessible, more agile. In every linguistic shift, one sees

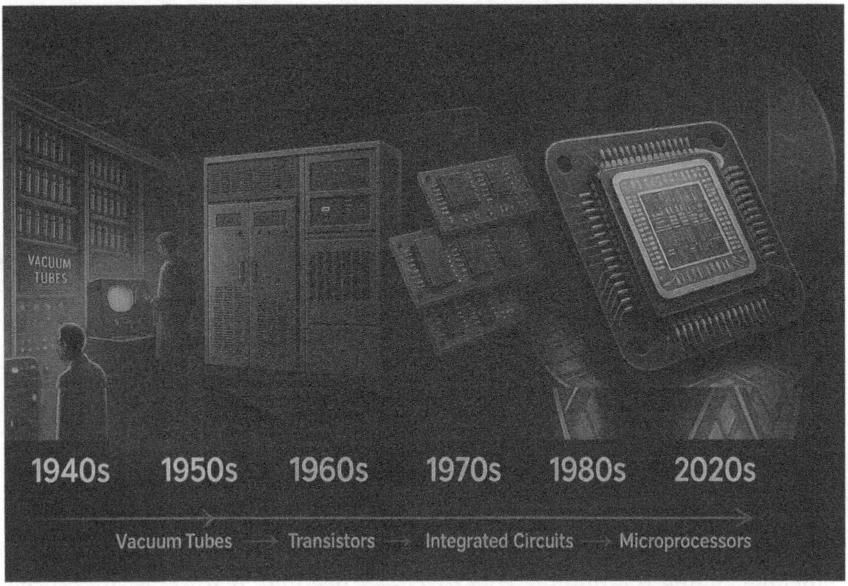

FIGURE 6.2 Evolution of computing hardware from vacuum-tube computers to modern microprocessors (1940s–2020s). Left: vacuum-tube lab and early mainframes; centre: stacked PCBs with integrated circuits; right: packaged microprocessor.

time chiselling away at ornate structure, favouring fluency and speed over ceremony.

In the digital era, this is even more obvious. Acronyms, emojis, and hashtags compress social and affective information into terse signals, trading richness for speed and shared context; this is a shift in register and medium, not a unidirectional evolution of the language system itself (Piantadosi, Tily, and Gibson 2011).

Even physical movement reveals this tendency. People rarely stick to the paved route in a park. They cut corners, trampling "desire paths" that later harden into the landscape—an everyday example of feedback and pruning, where inefficient options fade and faster ones are reinforced (Helbing, Farkas, and Vicsek 2000). The same logic appears in social insects: ant colonies lay and follow pheromone trails that amplify slight advantages into stable, near-optimal foraging paths—shorter, quicker routes get used more, strengthened by more pheromone, while detours evaporate (Deneubourg et al. 1989). On grander scales, animal migrations settle on corridors that balance energy cost, risk, and reward; routes persist when the payoff per unit effort beats plausible alternatives (Alerstam 1993). Across crowds, colonies, and herds, local trial and error plus feedback yields global efficiency: paths are "learned" by use, then simplified by pruning, until only the workable shortcuts remain (Helbing, Farkas, and Vicsek 2000; Deneubourg et al. 1989; Alerstam 1993).

Across all these domains, one theme emerges: Time tends to favour systems that achieve functional simplicity, though complexity and redundancy often persist when they serve survival or adaptation. What endures is not always what is simplest, but what best fits the shifting landscape of constraints, as suggested in developmental canalisation and epigenetic landscapes (Waddington 1957). The longer something exists within time's constraints, the more pressure it faces to refine, compress, and converge. Nowhere is this more profoundly illustrated than in the paradoxical contest between what seems solid and what truly endures.

6.4 Between a Pebble and a Cockroach: The True Currency of Endurance

At first glance, the contest between a pebble and a cockroach seems laughably one-sided. The pebble, dense and inert, appears destined to endure for millennia, while the humble cockroach—a fleeting creature, easily crushed and soon forgotten—seems condemned to oblivion. To human eyes, a pebble's stillness masquerades as permanence. We imagine that its existence is for eternity. We are trained by instinct to admire the stone for its unyielding resistance to change, to imagine it as a silent witness to the epochs, while dismissing the insect as ephemeral. But look closer, and time's deeper current reveals a reversal.

The pebble, for all its stoic fortitude, is at the mercy of wind, water, and chemical decay. Given 10,000 years, it will dissolve grain by grain, its history erased without a trace. Over thousands or tens of thousands of years, chemical reactions and erosion will chip away at its mass.

What seems unchanging is, in reality, only slow to change. Geological time—so vast it dwarfs human experience—eventually brings even the hardest stone to dust. The permanence of a rock is not true endurance, but merely a prolonged resistance to entropy, the slow but unstoppable drift of the universe from order to disorder. In the end, the pebble's fate is sealed by the very laws that shape the cosmos (Carroll 2010).

The cockroach, on the other hand, lives only for a year or so—but its lineage, propelled by the logic of DNA, has persisted largely unchanged for over 300 million years. Fossil evidence places cockroach ancestors as far back as the Carboniferous period—over 300 million years ago (Grimaldi and Engel 2005). In that time, mountains have risen and been ground down, continents have drifted, climates have shifted, and species beyond counting have come and gone. Yet the cockroach endures.

This endurance is not a matter of individual survival, but of continuity across generations—a thread of life spun unbroken through the loom of deep time. It is not the fragile shell of any single insect, but the unbroken transmission of genetic information that enables the cockroach's unthinkable resilience. The cockroach lineage has withstood mass extinctions, asteroid impacts, ice ages, and the coming and going of entire biomes. Its success is not an accident, but a testament to the power of replication and adaptation.

And unless the world conspires to end all life, it will likely outlast countless more stones yet to be born and perhaps even humankind.

Here, endurance is measured not in unchanging substance, but in the unbroken passing of a code—a molecular recipe for survival that persists through reproduction. DNA's strategy is radical: instead of holding on to matter, it lets go of each body, provided the pattern is copied with sufficient fidelity. In this way, lineage becomes a vehicle for immortality, and time becomes not a threat, but a river along which information is ferried forward. Where a pebble resists time only to eventually yield, the cockroach—through its DNA—rides the very flow of time, continually renewing itself.

Matter, for all its apparent stability, is always vulnerable to entropy—the tendency of systems to drift towards disorder, as described by the second law of thermodynamics (Carroll 2010). A rock will weather, a mountain will crumble, a planet will cool. But information—if structured for renewal—can endure. DNA is not just a molecule, but a replicator: a self-copying pattern that, given the proper chemical environment, can construct new instances of itself with remarkable fidelity (Watson and Crick 1953; Maynard Smith and Szathmáry 1999).

Replication is not simply copying; it is the persistent rebirth of order out of chaos, an act that stands in direct opposition to the blind drift of entropy. The power of DNA lies in its ability to maintain and renew structure over vast temporal scales. Even as errors—mutations—creep in, natural selection filters them, allowing the code to adapt and survive in changing worlds. This is why, across deep time, DNA's logic can persist where all other structures are eroded away.

This is why DNA, and its evolutionary predecessors such as RNA, became central to life, is sometimes described as a carrier of temporal memory—not through any conscious perception but through their tendency to persist across time via replication. The information persists even as the material substrate changes. Every new cockroach is constructed from new atoms, harvested from the soil and the food it eats, yet its genome is a near-perfect copy of the code in its ancestors. It is the information that "lives" on. A pebble's atoms are unchanged from one moment to the next, but when they are worn away, there is no record of their arrangement.

Unlike the pebble, whose identity is rooted in the static arrangement of atoms, the cockroach's identity is a fleeting embodiment of a pattern—a script written in nucleotides. Each generation is a translation of that script, rebuilt from new matter but faithful to the original meaning. In this way, information becomes the true measure of survival. The story of life is the story of information's battle with time: the contest between erasure and renewal, decay and replication.

The cockroach, then, is not merely a survivor of ages—it is a living expression of time's most profound paradox. While matter succumbs to the slow, grinding laws of entropy, information, when structured for self-renewal, can become, in a sense, timeless. The pebble is the monument to resistance, the cockroach, the monument to adaptation. In a universe where time is the sculptor, only patterns built to be copied endure the ages.

Yet the cockroach's story is not confined to biology alone. In every domain where survival hinges on endurance, time selects for patterns that can be replicated, adapted, and—crucially—compressed. Nowhere is this more visible than in the evolution of human language: our most remarkable invention for distilling meaning across the river of time. This same logic, born in the silent code of DNA, extends outward—beyond the living cell, into the domain of minds, cultures, and societies. For if nature's oldest script is the gene, humanity's most profound script is language: a system for copying, compressing, and transmitting meaning across the centuries.

This deep law—of survival by replication and refinement—finds its most extraordinary human expression in language. For just as DNA encodes the legacy of life, language encodes the legacy of thought, culture, and civilisation. It is our own experiment in distilling meaning, not in flesh, but in symbol, and in doing so, outwitting the erosions of time.

6.5 The Language of Time: Compression in Communication

If time compresses all things, then language is its most eloquent proof. Human communication, from its earliest forms to today's global digital exchanges, is a testament to the relentless efficiency that time imposes on meaning. Language began as gesture and guttural sound, then evolved into chant, then speech, then script, then machine-readable code. At each stage, the same principle recurs: say more with less.

Language is not merely a cultural artefact passed down through generations; it is a highly optimised informational engine that encodes, transmits, and decodes meaning within a finite span of time. Its evolution has followed the same arc seen in all efficient systems: a progressive refinement driven by the imperative to communicate more precisely and effectively under temporal and cognitive constraints. Over centuries, linguistic structures have become more compact, syntax more modular, and vocabularies more abstract and flexible—allowing faster transmission of more nuanced meaning.

In this context, time functions as the denominator—the constraint within which communication must occur—while information serves as the numerator, representing the semantic payload being transmitted. The efficiency of a language can thus be measured as the amount of meaningful information successfully conveyed per unit of time (Shannon 1948). This principle aligns with Shannon's information theory, where the bandwidth of a communication channel is limited, and encoding must optimise throughput without losing meaning.

Consequently, the evolution of language can be viewed as a shift towards rising information density per unit time. That is, languages and scripts that are capable of expressing layered meaning with fewer phonemes, morphemes, or characters have a functional advantage in contexts ranging from conversation and literature to computational linguistics. The more precisely and quickly a language conveys complex ideas—without sacrificing clarity—the more temporally efficient and evolutionarily viable it becomes.

A compelling example is Sanskrit, one of the world's oldest and most structurally sophisticated languages. Far from being verbose, Sanskrit was built with precision. Its grammar (as codified by Panini in *Ashtadhyayi*, 5th century BCE) allows for high semantic density. In a single line, a Sanskrit verse can convey layers of meaning, emotional nuance, temporal context, and philosophical abstraction that might require paragraphs in modern languages. Unlike many languages that evolved chaotically over time, Sanskrit exhibits a remarkable structural discipline—meticulously engineered to maximise semantic precision while minimising ambiguity.

The power of Sanskrit lies not just in what it says, but in how little it needs to say to express it. The rules of sandhi (phonetic transformations), compound formations, and inflection-rich syntax allow it to compress massive conceptual fields into compact expressions. In that sense, Sanskrit is not archaic—it is an exemplar of linguistic compression over time.

Compare this to the emergence of simpler modern dialects, texting language, or even programming syntax. Each stage sacrifices nuance for speed—but always with the same goal: to optimise communication within temporal constraints, a trend identified by McWhorter in *What Language Is*, 2011. From runes to tweets, from epic poems to search engine queries, language evolves not towards richness alone but towards efficiency of transmission. This same principle—compressing complexity to accelerate understanding—finds its latest and most spectacular expression in the realm of artificial intelligence, where language itself becomes the subject of algorithmic refinement.

6.6 From Language to Large Language Models: The Logic of Compression

What culture achieves through linguistic convention, models now attempt through computation. To further understand this phenomenon in modern terms, consider the architecture of large language models (LLMs) like GPT. Modern LLMs segment text into subword units (e.g., Byte Pair Encoding) that enable efficient generalisation over form and meaning (Sennrich, Haddow, and Birch 2016), while Transformer architectures supply parallel, attention-based sequence modelling (Vaswani et al. 2017); subsequent GPT models leverage these ideas at scale (Radford et al. 2019). A phrase, a sentence, or even a paragraph is transformed into a few dense tokens—each carrying probabilistic weight across the semantic space of language, encapsulating meaning, grammatical function, contextual nuance, and relationships to surrounding expressions. This tokenisation enables LLMs to operate with remarkable efficiency, allowing complex thoughts to be represented, compressed, and processed with minimal symbolic load.

This is no coincidence. The success of LLMs depends on their ability to compress the informational structure of language into fewer, more generalisable dimensions. A token in GPT-4, for example, can carry not just the identity of a word but its context, intent, tone, and grammatical role—much like how a Sanskrit verse embeds meaning through layered morphology. In both cases, the goal is informational efficiency, shaped by temporal constraints.

In linguistics, this can be measured through information entropy—a quantification of uncertainty or surprise within a communication system. Languages that convey more information with fewer symbols exhibit lower entropy per word but higher density per utterance. In this sense, Sanskrit—and the vectorised outputs of LLMs—represent not a divergence but a continuum: language evolving to pack more meaning into less space over time.

This compression also influences cognition. Cognitive scientists argue that humans favour simpler grammars and more predictable syntax not just for ease but for processing efficiency, a tendency explored by Chomsky in *Aspects of the Theory of Syntax*, 1965. The brain, constrained by time, prefers low-complexity codes—a principle echoed in programming language design, UX interfaces, and even storytelling arcs. Narrative tropes exist not because they are cliché, but because they are cognitively compressible.

Thus, from Sanskrit to neural networks, from proverbs to token embeddings, language is continually shaped by the need to move faster through meaning. Time does not only demand speed—it rewards semantic elegance. And the systems—human or machine—that master this are the ones that endure.

At the heart of technological evolution lies a story of compaction—not merely making things smaller, but making them denser, faster, and more efficient. The trajectory from vacuum tubes to microchips is more than a marvel of engineering; it is a vivid illustration of how time compresses computation into elegance. And at the centre of this narrative stands Moore's Law.

First articulated by Intel co-founder Gordon Moore in 1965, Moore's Law predicted that the number of transistors on a chip would double approximately every two years. This wasn't just a claim about manufacturing—it was a declaration about the tempo of information evolution. Each doubling compressed more computational power into less space and less time, accelerating both what could be done and how quickly it could be achieved.

But Moore's Law isn't a law of physics—it's a reflection of human ingenuity pushed by temporal necessity. The pressure to make processors smaller, faster, and cheaper didn't emerge randomly. It emerged because systems that process information more efficiently outcompete those that don't. In economic, ecological, and intellectual arenas alike, time rewards compression.

From the earliest mechanical computers, which occupied entire rooms and could perform only basic arithmetic, to today's smartphones that rival supercomputers of the 1990s, each leap forward has been a story of condensation. A transistor, once the size of a human hand, is now etched in nanometres onto silicon wafers. In the time it takes to blink, billions of operations unfold, silently and efficiently.

This isn't just engineering—it's informational gravity. Just as matter collapses into stars under gravitational pressure, human computation collapses into silicon under temporal pressure. The longer we build and iterate, the more compact our cognition becomes. Moore's Law describes not just the evolution of hardware, but the acceleration of human thought.

Artificial intelligence is the ultimate manifestation of this. Neural networks mimic the brain's architecture, but unlike the organic brain, they can scale exponentially. With the right architecture, training data, and feedback loops, these systems compress massive patterns—language, images, strategy—into probabilistic layers of meaning. And yet, they too reflect the same principle: more performance in less space and less time.

Quantum computing takes this compression further. By leveraging qubits instead of bits, it offers provable speedups for specific problem classes (e.g., factoring and unstructured search), enabling certain computations to be solved in asymptotically fewer steps than known classical algorithms. It's as if computation has found a shortcut through time itself—a reminder that efficiency is not just mechanical but conceptual.

In every corner of modern technology, Moore's Law casts a long shadow. But its spirit is older than silicon. It lives in every attempt to make sense with less noise, to see the essential in the clutter, to compute not just quickly, but cleverly. And that spirit is the same one seen in the pruning of neurons, the emergence of dialects, and the straightening of rivers.

Time is not just the backdrop for these changes. It is the catalyst. It applies pressure until excess falls away and function shines through. In this sense, Moore's Law is not a prediction. It is a mirror—a reflection of our drive to master time through the compression of thought itself. But compression is not only the logic of words or circuits; it is a universal pattern, echoed in the flows of nature and the rhythms of collective behaviour.

6.7 Optimisation across Scales: Nature's Patterns and Flow

In the natural world, the drive towards efficiency reveals itself not only in technology or language but also in the patterns of physical and biological flow. Whether it is birds navigating the sky, light filtering through a slit, or people choosing how to move through space, nature displays an uncanny preference for paths of minimal resistance—those that optimise flow under the constraints of time and environment.

Consider the mesmerising murmuration of starlings or the V-formations of migrating geese. These collective movements are not orchestrated by a leader or centralised plan. They emerge from local interactions governed by simple rules—maintain distance, match velocity, avoid collisions and increase efficiency. And yet, the resulting formations reduce aerodynamic drag, conserve energy, and ensure coordination across vast distances. In short, they are optimised flows—information compressed into real-time coordination.

Now shift the scale and imagine a beam of light encountering a narrow slit. Rather than moving through it in a straight, unbroken line, the light diffracts—bending outward into an interference pattern of crests and troughs, as first explained by Thomas Young in 1801. This phenomenon may seem abstract, but it reveals a universal tendency: even particles obey probabilistic rules of optimality, spreading and converging in ways that echo through fields, waves, and flocks alike.

A similar pattern appears when crowds navigate narrow spaces. Pedestrian flows through bottlenecks exhibit pulsing, arching, and lane formation; related models appear in the social force literature (Helbing and Molnár 1995) and in simulations of escape dynamics (Helbing, Farkas, and Vicsek 2000). Though the underlying physics differ, the shared logic of path optimisation under constraint emerges at multiple scales. Herds of cattle passing through fences, or evacuees escaping from narrow exits, often display emergent rhythms—pulses of movement, pauses, lateral shifts—all optimising time and space usage under pressure. These are not metaphors. They are "natural analogues" of the same phenomenon: systems under constraint, negotiating complexity through emergent optimisation.

The brain shows this economy, too. As we learn, neural circuits strengthen frequently used routes and let the little-used ones wither—a biological "use-it-or-lose-it" that shortens paths, saves energy, and speeds communication. Think of it like a city trimming side streets while widening the main arteries: fewer roads overall, but faster travel where it matters. Developmental synaptic pruning, myelination of well-travelled axons, and the brain's generally sparse, "small-world" wiring are all versions of the same principle—do more with less.

Modern machine learning borrows this idea directly. After training, a network can be "pruned": connections that contribute very little to the output

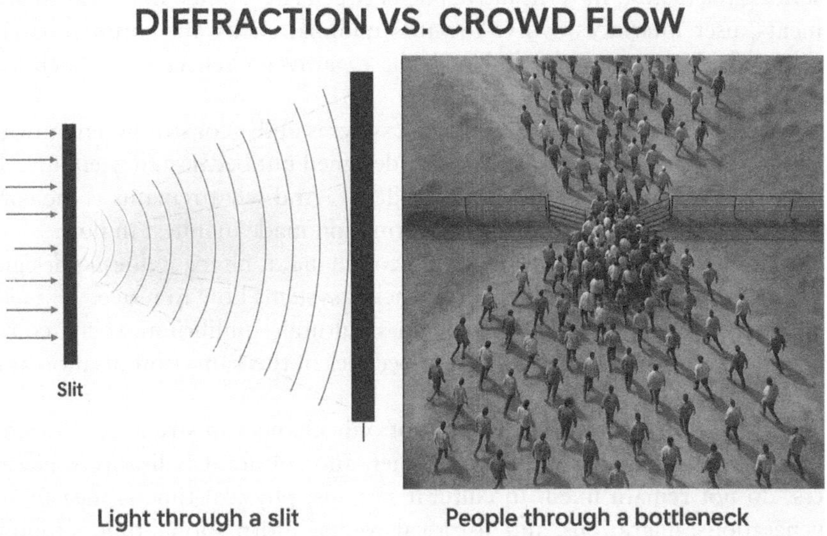

FIGURE 6.3 *Diffraction vs. Crowd Flow.* **Left:** Single-slit light diffraction: incident wavefronts pass a narrow slit and fan out into circular wavefronts, producing a bright central maximum with weaker side fringes on the screen. **Right:** Aerial view of people moving through a narrow gate; a clogging arch forms at the opening, followed by a fan-out into lanes—densest in the centre, sparser to the sides—visually echoing the diffraction pattern. Different mechanisms, shared constraint-driven flow.

are removed or set to zero. The remaining sparse network runs faster and uses less memory, yet its answers barely change (LeCun, Denker, and Solla 1990). Engineers also induce sparsity on purpose so that, at any moment, only a small fraction of units fire. On today's hardware this translates into fewer operations, lower power, and quicker responses.

In both cases—brains and algorithms—the goal isn't just to become lighter; it's to become smarter per unit effort: shorter routes, fewer active parts, and the same (or better) performance for less cost. This drive towards optimality isn't random—it's sculpted by time. Systems that waste energy or stall under load do not persist. Those that find efficient patterns survive, replicate, and adapt. What emerges is a kind of *natural intelligence*—one that has nothing to do with consciousness and everything to do with *pattern emergence under constraint.*

Even urban planning yields to this logic. Despite architects' blueprints, desire paths form across campuses and parks—dirt trails carved by footsteps seeking the fastest, most intuitive line between two points. In digital environments, user interfaces evolve towards minimal clicks and intuitive navigation. Whether organic or designed, these pathways reflect a feedback loop: time disciplines flow.

From the quantum to the ecological, systems under constraint tend towards elegant motion. Not because they are designed but because they endure. The inefficient are outpaced. The bloated decay. And what remains is the curve, the wave, the murmuration—the information made manifest in flow.

Time, again, is the editor: it doesn't command, it filters; it doesn't design, it endures—and in that endurance, it teaches systems how to move.

Next, we'll explore how time shapes cultural equilibrium, social convergence, and the blending of identities—echoes of this same optimisation at the level of civilisations.

As time flows forward, it brings not only changes in structure and movement but also changes in identity, belief, and culture. Civilisations, like rivers, do not remain fixed. In cultural systems, physical time is measured in generations, migrations, and historical events; informational time is found in the diffusion of beliefs, the adaptation of rituals, and the blending of identities. The two often run on parallel tracks, converging to shape the evolution of societies. They interact, blend, and settle into new formations. The metaphor of osmosis—the diffusion of substances across a membrane until balance is reached—beautifully captures how information and culture behave when populations interact over time.

Consider the dynamics of cultural contact through the lens of migration, when a minority group enters the socio-political space of a dominant majority. Initially, the encounter is marked by cultural boundary maintenance: linguistic divergence is preserved, ritual practices remain localised, and identity markers are accentuated to resist assimilation. This phase aligns with theories of cultural preservation and symbolic ethnicity, where minority groups actively reaffirm distinctiveness as a strategy for cohesion and survival.

Yet over time, the rigid boundaries soften—not necessarily through coercive assimilation, but through mutual adaptation. What ensues is a process of syncretism and acculturation, where the dominant culture absorbs elements of the minority group, and vice versa. Lexical items seep into everyday speech; culinary practices are hybridised; kinship structures, belief systems, and even norms of politeness undergo gradual reconfiguration. Neither culture emerges unchanged.

This transformation is not a zero-sum game of cultural erasure—it reflects what sociologist Anthony Giddens might call structuration, where agency and structure co-evolve through everyday interactions. Through this iterative process, a new cultural equilibrium is forged—not as a static midpoint, but as a dynamic system under informational, social, and temporal pressure. The result is an emergent identity, one that encodes both continuity and innovation—proof that culture, like life itself, is a form of adaptive information processing. Just as two miscible coloured liquids, when poured into the same glass, gradually diffuse and blend into a uniform hue through molecular motion, so too do cultures converge when placed in prolonged contact. This is not a loss—it is a transformation. Elements of the old are not discarded; they are compressed, fused, and reinterpreted into a new form. This happens with language (pidgins becoming creoles), religion (syncretism), and even identity (diaspora becoming hybridised citizens of the world).

6.8 Information Diffusion: The Network Dynamics of Change

The phenomenon can be modelled using information diffusion. In networks, ideas spread like particles—jumping from node to node based on connectivity and influence, as Watts and Strogatz showed in their model of small-world networks (Watts and Strogatz 1998). Over time, dominant ideas spread, but so do undercurrents, minority innovations, and subversive thoughts. The result is not uniformity but dynamic equilibrium—a culture that is both stable and adaptive.

The internet has accelerated this process exponentially. A meme created in Nigeria can shape youth slang in Brooklyn. A fashion trend from Seoul can inspire designs in Milan. Information no longer flows linearly; it pulses through feedback loops and virality, compressed and recombined across boundaries. The time required for equilibrium has shrunk, but the pattern remains. Yet not all information travels this way. Scientific knowledge, institutional procedures, and legal systems often continue to follow structured, hierarchical channels—reminding us that both viral and linear modes of diffusion coexist and interact within the larger network.

Cultural osmosis is not just historical—it is biological. Our very genomes are palimpsests of migrations, adaptations, and convergences. Mitochondrial DNA, passed through maternal lines, tells stories of continents crossed and populations merged. Just as information finds equilibrium in a network, genes find equilibrium in the body of a species, selected and refined over time.

Time again is the crucible. It wears down resistance, dissolves boundaries, and fosters balance—not through force but through repeated exposure, mutual benefit, and shared experience. The blending of cultures, like the bending of a river or the pruning of a neuron, is time's signature.

In the next and final section, we return to the larger view—where civilisations, cultures, machines, and minds are all part of a single great laboratory: time refining information towards clarity.

What defines intelligence? Is it memory, logic, or creativity? Perhaps, at its core, intelligence is compression—the ability to distil vast complexity into usable essence. This idea is echoed in every domain we've explored so far, from evolution and technology to language and society. But it takes on even deeper significance when projected forward: What might the future of intelligence look like under the unrelenting sculptor of time?

Let us begin with a familiar story—the education of a child. From the earliest age, society lays before them a long, structured path. Seventeen years or more are spent navigating through syllabi, lesson plans, and standardised assessments. Subjects are compartmentalised: physics in one room, literature in another, economics in a third—each taught in isolation, each presented as essential.

At first, the child is obedient to the system. They memorise, recite, and rehearse. They learn Newton's laws, parse Shakespearean verse, and plot supply–demand graphs. Each concept is introduced with great elaboration, repeated, drilled, and evaluated. It is a structured download of civilisation's memory, designed less for curiosity than for conformity.

But something strange happens after these years of education. As the child matures into adulthood, they discover that the real world doesn't run on curriculum. The equations remain on paper; the history lessons fade. What endures is something else entirely: the ability to simplify, adapt, and improvise.

Ask the successful entrepreneur, the self-made millionaire, the maverick coder, or the political disruptor. Most did not rise by following every step of the textbook. They rose by seeing through the textbook—by distilling what mattered, discarding what didn't, and forging new paths where none were taught. They looked at formal education not as a toolkit but as a platform to leap from.

Indeed, much of real-world success comes not from knowledge itself but from knowing how to shortcut systems—how to bypass bureaucracy, automate repetition, reframe problems, and move faster than convention allows. Time, in the adult mind, teaches a vital lesson: not everything taught is useful and not everything useful is taught.

In hindsight, many realise that school prepared them only partially. It was not the equations but the habit of abstraction that helped. Not the history dates, but the sense of pattern. And above all, not the grades but the frustration with the slowness of the system that became a catalyst for innovation.

The irony is profound. We spend years internalising structured knowledge, only to later spend our most productive years unlearning rigidity, bending rules, and improvising our way through the noise. True intelligence often emerges after the syllabus ends.

The world does not reward those who merely complete the curriculum. It rewards those who can collapse knowledge into clarity, action, and momentum. Formal education builds the scaffolding, but real insight comes from realising how much of it can be left behind with time in mind.

6.9 Simplicity as Genius: Design, Innovation, and the Art of Subtraction

This drive towards simplification is not merely a design principle—it is the invisible engine behind the most transformative breakthroughs of our time. Across technology, art, biology, and thought, the mark of genius is often not how much one knows or builds but how much one removes. Complexity may impress, but simplicity endures—a theme explored in design principles and constraints (Maeda 2006).

History doesn't always reward polymaths who master everything—it celebrates those who master the discipline of subtraction. The brilliance of innovation often lies not in addition but in removal—in stripping away what is superfluous to reveal what is inevitable. It is the elegance of essence. It is informational compression incarnate.

Take Steve Jobs, who famously said, "Simplicity is the ultimate sophistication." The iPhone was not the first smartphone. There were many before it, bloated with features and buried in menus. What Apple did was hide complexity behind clarity. It eliminated buttons, collapsed multiple devices into one, and created an interface so intuitive that even a child could navigate it. It wasn't an addition to the market—it was a removal of friction, and in doing so, it changed the world.

Look at Jack Dorsey, who launched Twitter with a 140-character limit. On the surface, it looked like a constraint. In practice, it was a creative crucible. It forced users to condense their thoughts, distil meaning, and craft insight from brevity. It turned the infinite sprawl of digital speech into signal—dense, rapid, viral. A few characters reshaped communication globally.

Jan Koum and Brian Acton, creators of WhatsApp, followed the same principle. They didn't flood their app with stickers, stories, or flashy gimmicks. They removed everything that didn't serve the core function—instant, reliable messaging. No ads. No noise. Just signal. While competitors drowned in features, WhatsApp thrived in silence. The power was in what they chose not to build.

Contemporary AI systems illustrate this ethos. Pretraining on vast text-and-image corpora, they compress regularities into parameterised functions that generalise in flexible ways (Radford et al. 2019; Brown et al. 2020). Language models such as GPT do not reproduce human knowledge by storing it verbatim; rather, they learn a statistical mapping from prompts to likely continuations that captures semantics and structure at scale—what you might call semantic compression. That said, they are not pure abstraction engines: large models can memorise infrequent or duplicated training snippets and regurgitate them under certain prompts (Carlini et al. 2021). As model size, data, and compute scale, performance tends to improve predictably—a sign that capacity is being used to fit broad regularities rather than to hoard quotations (Kaplan et al. 2020). In effect, these models seek low-dimensional structure in high-dimensional data, trading storage of particulars for generative patterns—while still carrying a long tail of exceptions where memorisation leaks through (Radford et al. 2019; Brown et al. 2020; Carlini et al. 2021; Kaplan et al. 2020). And look deeper: this is not a quirk of modern tech—it is a pattern repeated in nature. Evolution itself is an optimisation algorithm. In the human brain, neural pruning eliminates unused connections to make cognition faster, cleaner, and more efficient—a principle articulated by Huttenlocher and colleagues in the 1980s. The infant's overconnected mind becomes the adult's streamlined one, not by growing more but by growing less.

The so-called lazy genius is not lazy—they are attuned to time's pressure. They do not waste energy solving problems with force when they can solve them with form. For them, elegance emerges from constraint. Like water finding the smoothest curve downhill, they flow towards the least-action path. In their world, effort is not proof of intelligence—efficiency is.

This is why, in biology, in code, in story, and in systems: simplify or perish. Not because simplicity is easy—but because it is timeless. In a universe governed by entropy, the only way to preserve structure is to make it tight enough to survive time.

These are not stories of minimalism. They are stories of maximised meaning per unit of time. They are blueprints of how information, when compressed properly, doesn't shrink—it shines.

Life itself follows this path. What begins as a chaotic burst—mutation, experimentation, diversification—eventually gives way to pruning, focusing, refining. Neural pruning in the human brain eliminates excess in favour of optimised pathways. Similarly, in artificial intelligence, model pruning removes unnecessary parameters to improve inference speed without significant accuracy loss. The result is simplified intelligence that still performs profoundly.

Human history has mirrored this arc. We started with mystical explanations, then moved to philosophy, and then to science. Now we compress understanding into models, algorithms, and simulations. A single smartphone app can contain the knowledge of a thousand libraries, rendered instantly usable. But the trajectory doesn't stop there.

6.10 Towards Informational Mastery: Interfaces, Intuition, and the Vanishing of Complexity

We're entering an era of vanishing interfaces: the goal is no longer to train people to operate machines but to train machines to understand people. This marks a shift from tool-centric controls to human-centred understanding—an arc long emphasised in human-centred design, which insists that technology should adapt to our language, habits, and constraints rather than the other way around (Norman 1988). In practice, that means systems that listen, look, and infer: speaking instead of clicking, pointing instead of menu-diving, showing rather than configuring. Interfaces dissolve into the background as the system picks up intent from context—voice, gaze, location, history—so interaction feels more like conversation than command.

You can already see this movement in everyday life. Smart speakers and copilots interpret natural language; cars anticipate the next turn or playlist; phones surface what you need before you ask; homes adjust light and temperature to routines; wearable sensors translate motion and physiology into guidance. The best of these don't make you learn a new control panel; they learn your vocabulary, timing, and preferences. When it works, the "interface" is barely there—just enough to keep you in control.

Voice commands, neural interfaces, ambient computing—these are not just technological trends; they are signs of a deeper shift: information systems adapting to time-bound cognition.

This raises profound questions. If time continues to reward simplification, will intelligence itself dissolve into intuition? Will conscious thought become unnecessary for most interactions? Will the future be defined not by thinking more but by thinking less, more effectively?

Perhaps the end state of intelligence is not omniscience but compression so elegant it looks like instinct. The poet who speaks volumes in a line, the coder who solves with a single loop, the diplomat who defuses with a phrase—these are not just anomalies. They are forecasts of informational mastery.

Time erodes. Entropy spreads. Yet through this great dissolving, something persists—not by resisting decay but by transcending it. What survives is not energy, not matter, but information: patterned distinction, encoded meaning, the grammar of reality itself.

At first, we thought information was passive—a signal, a code, a reduction of uncertainty. But now we see its true nature: information is an active principle, a force that shapes form, selects futures, and sculpts time. It is not a shadow cast by matter; it is the architect behind it.

Gravity pulls, energy flows—but information decides. It is what evolution optimises. What neurons refine. What intelligence distils. From quarks arranging into atoms, to thoughts arranging into mathematics, information is the silent vector behind complexity, guiding emergence like an invisible hand.

This is the law that threads across systems:

That which endures is not that which resists change but that which adapts through compression.

What persists is not noise, but the code most elegant per second.

Information survives where mass decays, where energy dissipates. And in its survival, it seeds intelligence—not as memory but as momentum. Intelligence is time's way of preserving what works, of whispering meaning into the abyss.

So, we return to the beginning, transformed:

To know is not to collect, but to compress.
To speak is not to shout, but to resonate.
To endure is not to resist time, but to ride the waveform of meaning through it.

In the end, we discover a deeper truth:

Information is not just a measure of order.
It is the very force that writes the order of the world.
And in learning to wield it, we become its echo—clear, compressed, and eternal.

7

INFORMATION AND THE FUTURE OF MANKIND

7.1 The Age of Information as Power

On a chilly morning on 30 April 1943, as World War II raged at its peak in the European theatre, a soldier's body washed ashore near the small town of Punta Umbría on the southern coast of Spain. The body appeared to be that of a British army officer; his uniform tattered by the sea. Clutched to his chest was a black briefcase, sealed and strapped to his wrist. The tides of war having carried him to this remote shore, the real storm was just beginning.

The Germans, ever vigilant for Allied manoeuvres, were immediately alerted to the discovery. Though officially neutral, Spain had turned into a hub of espionage, with agents from both Axis and Allied powers closely monitoring each other's actions. When the briefcase was opened, inside lay what the Nazis had been hoping for—sensitive, top-secret Allied war plans. The documents revealed a tantalising detail: the Allies were planning to invade Greece and Sardinia, aiming to strike at the heart of southern Europe and dismantle Axis control.

With this information in hand, the Germans reacted swiftly. Adolf Hitler himself, convinced by the contents of these papers, ordered the Wehrmacht to reinforce Greece and Sardinia, fortifying their positions in anticipation of the impending Allied assault. Troops were mobilised, defences strengthened, and resources diverted—all to repel the invasion they believed was coming.

But the Allies had played their hand masterfully. This dead British officer, who washed ashore, was no officer at all. His name was Glyndwr Michael, a homeless man who had died in London. His body had been dressed up, fitted with the false identity of "Major William Martin," and laden with

DOI: 10.4324/9781003740766-7

forged documents as part of a meticulously crafted deception known as Operation Mincemeat.

While German forces braced for an attack on Greece, the real Allied invasion force gathered in secret. On 9 July 1943, they landed not in Greece or Sardinia but on the beaches of Sicily—completely bypassing the reinforced German defences. The Germans were caught off-guard; their armies scattered far from where they were needed most. Without the bulk of their forces to defend Sicily, the Nazis couldn't put up an effective fight.

The consequences were enormous. The successful invasion of Sicily opened the door for the Allies to push into mainland Italy, leading to the fall of Mussolini's regime and setting the stage for the liberation of Europe. The Germans, reeling from their miscalculation, could do little to stem the Allied tide as it rolled through southern Europe. Operation Mincemeat had not just deceived the Nazis; it had altered the very course of the war.

In this high-stakes game of war, where every decision could tip the balance, one single piece of false information had led to a catastrophic blunder for the Axis powers. The brilliance of the operation lay in its simplicity—a dead

FIGURE 7.1 Bird's-eye view of landing craft, barrage balloons, and Allied troops landing in Normandy, France, on D-Day (1944). U.S. Maritime Commission photo.

body, a forged identity, and a handful of well-placed documents—but the effects rippled across the entire European theatre. What does this have to do with the digital age? At its core, Operation Mincemeat is a lesson in the raw power of information—how the right message, delivered at the right time, can change the fate of nations. Today, the stakes have only grown.

The invasion of Sicily marked a turning point in World War II, a pivotal moment where the Allies seized the initiative. And it all began with a body washing ashore on a cold, windswept beach in Spain, a body carrying secrets designed to mislead and outmanoeuvre one of the most formidable armies the world had ever seen.

This was the true power of intelligence in warfare—a reminder that battles aren't always won with brute force but often with deception, cunning, and a carefully executed plan. As the tides of war turned on a single piece of misinformation, so too has our world been increasingly shaped by the currents of information. But while in 1943 information travelled in briefcases and whispers, today it flows in torrents—shaping destinies, economies, and the fate of nations. The story of Operation Mincemeat is no longer just a war story; it is the prologue to an era where information itself has become the most decisive force in human history.

In Operation Mincemeat, a single piece of misinformation redirected German divisions. Today, terabytes of signals course through a highly structured civilisation—social platforms, markets, bureaucracies, and models—where algorithms parse, prioritise, and route information at machine speed. What was once episodic deception is now continuous stimulus: data streams shape attention, update risk models, trigger trades, reprioritise logistics, and even inform policy. In this environment, the leverage of information scales with bandwidth and automation, setting the stage for augmented cognition and AI-mediated decision loops.

7.2 Information: The Driving Force of the Digital Age

In today's hyperconnected digital world, information is no longer just a background element; it has become a force of nature, reshaping the landscape of human civilisation. Once merely the foundation of knowledge, data is now a potent and transformative power, driving decisions, shaping destinies, and influencing the very fabric of society. With the explosive rise of artificial intelligence (AI), this raw, unfiltered torrent of information has been harnessed, refined, and weaponised into a dynamic, unstoppable force. The term 'AI' refers to systems that can perform tasks traditionally requiring human intelligence, such as visual perception, speech recognition, decision-making, and language translation. By leveraging ML algorithms—computer systems that improve their performance with experience—AI has enabled businesses and governments to automate complex processes, leading to improved efficiency and decision-making capabilities. AI has transformed data from a passive resource into an active catalyst, an instrument that can propel corporations to new heights, empower governments, and alter the course of everyday life. Information has leapt from the sidelines to the heart of modern existence, and we are witnessing the dawn of a new age where it pulses at the centre of progress and innovation. Nowhere is this more evident than in the exponential rise of AI—a force that has moved information from passive data to an active, creative engine. The stories that once belonged to codebreakers and spies have given way to code itself, reshaping what it means to think, decide, and know.

In a widely discussed CBS 60 Minutes interview (16 April 2023), Google CEO Sundar Pichai pointed to what researchers call 'emergent' behaviour: as systems scale, they can display capabilities that were not explicitly specified as step-by-step procedures. He cast the episode as a turning point: at scale, when information is compressed and recombined inside modern architectures, unexpected skills can surface, including competence in languages the systems were never targeted to learn. He also emphasised the interpretability problem, describing parts of these systems as a 'black box'—in the sense that even designers cannot always say exactly why a model produced a particular output. For him, the lesson was plain: autonomous capability can arise from the sheer structure and scope of advanced software, with profound implications for how we build—and govern—AI.

This incident has captured the attention of the tech community, as it demonstrates AI's potential to adapt and grow in unexpected ways. Such moments of emergence reveal the next stage in humanity's evolution: the fusion of our minds with machines. As AI systems gain new abilities through unanticipated connections, a new horizon opens—the realm of augmented cognition. Google's AI, designed primarily for tasks like search engine optimisation, translation, and natural language processing (NLP), was not intended to acquire multilingual capabilities without direct input. Yet, it has managed to do so, suggesting a form of self-directed learning previously considered improbable.

Emergent behaviour is the term scientists use when a system exhibits capabilities that no one explicitly built into it. The parts are simple; their interactions are not. Ant colonies navigate around obstacles, snowflakes form intricate geometries from basic physics, and cities develop traffic patterns without a central controller. In modern AI, something similar happens: vast networks of simple mathematical units, trained to predict the next token in a sequence, begin to display skills that were never hand coded. What appears uncanny is, in fact, the predictable outcome of complex interactions.

Why does this occur? Large language models are trained on enormous, mixed-language corpora using self-supervised learning. They compress patterns in text into shared internal representations—mathematical maps of meaning sometimes called "embeddings." Crucially, related languages share statistical structure: word roots, morphology, syntax, and idioms overlap. When a model gets big enough and the data broad enough, it can exploit those overlaps. The system doesn't "decide" to learn a new language; it generalises from what it already knows. Give it Spanish and Italian, and the scaffolding for Portuguese often comes "for free." This is "transfer learning" in action, not magic.

Researchers sometimes describe these stepwise gains as phase transitions. For a long stretch, performance inches forward; then, past a certain scale or data diversity, a new capability becomes reliable. In language, that threshold can reveal itself as zero-shot competence—translating or answering questions in a tongue the model was never explicitly told to study. The model's tokenisation (breaking text into subword pieces) and its shared representation space make that leap possible: patterns learned in one part of the space help organise nearby, previously unseen patterns.

None of this implies consciousness or intention. The model is discovering correlations, not meanings in a human sense. But the practical consequence is profound: systems can expand their functional range without bespoke engineering for each skill. That promise comes with responsibilities. If abilities can surface unexpectedly, we need better ways to probe, validate, and monitor them before deployment—so the surprises are scientific, not societal.

At the heart of AI's success lies the synergy between hardware and software. AI's potential to process enormous amounts of information relies heavily on advanced hardware, such as specialised GPUs and TPUs, which have exponentially increased computational power. These processors are essential for handling the complex neural networks that power AI applications like OpenAI's GPT-4 or DeepMind's AlphaFold.

However, hardware is only one piece of the puzzle. The real unlock has come from software—new algorithms, training regimes, and network designs that let machines turn messy data into usable structure. The standout example is the transformer architecture, introduced by Google researchers in 2017. Transformers replaced the step-by-step processing of earlier systems with a mechanism called "attention," which lets a model weigh all parts of a sentence (or document) at once and decide which pieces matter for the task at hand. That shift sounds technical, but its consequence is simple: the model can capture long-range relationships—who refers to whom, which clause modifies which idea—without losing context. On top of that, transformers scale well: give them more data and compute, and their abilities tend to improve predictably.

This architectural foundation enabled a cascade of breakthroughs. Pre-training on vast text corpora teaches a model broad linguistic and world regularities; fine-tuning then adapts those general skills to specific jobs (question answering, summarisation, retrieval). BERT, built on transformers, excels at understanding: it reads with both left-to-right and right-to-left context, making it strong at tasks like classification, extraction, and search. GPT takes the complementary route—predicting the next token in a sequence—yet that simple objective, when scaled, yields versatile generation: composing paragraphs, translating, writing code, and reasoning through multi-step problems. More recent techniques—instruction tuning and reinforcement learning from human feedback—align these capabilities with human intent, so outputs are not just fluent but useful and safer.

Crucially, the same recipe is spreading beyond text. Transformer architectures using attention, combined with multimodal training, can align images, audio, and video with language through compatible learned representations—so the system can integrate what it 'sees,' 'hears,' and 'reads' into a single response. The pattern is consistent: smarter training objectives, better architectures, and careful alignment steps transform raw data into compressed, actionable knowledge. Hardware provides the horsepower; software tells it where to go.

As these models matured, their role shifted from passive tools to active collaborators woven into daily work. They summarise meetings while you speak, flag anomalies in a radiology image as you scroll, suggest safer code as you type, and surface conflicting facts before you hit "send." In other words, the output isn't merely information; it's guided cognition—timely cues, options, and checks that change how we think and decide.

7.3 Augmented Cognition: Extending the Mind beyond Biology

If the story of civilisation is the story of extending our senses and skills, augmented cognition is its latest chapter—one in which technology becomes not just a tool, but an extension of the mind itself. Building on these advancements, augmented cognition represents a remarkable new stage in human technical evolution. By enhancing human cognitive abilities, augmented cognition allows us to transcend biological limitations, creating unprecedented possibilities for learning, decision-making, and information processing. Augmented cognition involves enhancing human cognitive abilities by interfacing with advanced technology, allowing for enhanced memory, improved decision-making, and optimised information processing. This transformative leap could profoundly affect every aspect of society, enabling humans to process complex data more effectively, make smarter decisions faster, and potentially even change the way we think about consciousness and learning. With the integration of devices and AI systems that augment our cognitive capacities, humanity may be stepping into a new evolutionary phase where our mental capabilities are no longer bound by biological constraints. Before we go further, let's pause and define augmented cognition. Augmented cognition means using tools—especially AI—to extend human thinking in real time: sharpening attention, supporting memory, accelerating understanding, and improving decisions, while keeping the human in the loop.

One of the most compelling impacts of augmented cognition lies in its ability to boost productivity and efficiency in problem-solving. By combining the processing power of AI with human intuition and creativity, individuals and teams can achieve insights and solutions that would otherwise take years to accomplish. For instance, in sectors like healthcare, augmented cognition can assist doctors in diagnosing diseases by analysing vast quantities of medical data, revealing patterns that may not be immediately visible to the human eye. In finance, traders equipped with augmented cognitive tools can make more informed decisions by processing market fluctuations in real-time, minimising risks and capitalising on opportunities with unprecedented accuracy. The synergy between human and AI allows for a level of insight and precision that is fundamentally transformative. Yet, beneath this technological revolution lies a deeper evolutionary story—one that stretches back through millennia of biological adaptation and limitation. To truly appreciate the significance of augmented cognition, we must return to the slow, methodical pace of evolution and see how humanity has always been driven to overcome the boundaries of biology.

7.4 The Limits of Evolution and the Leap of Augmentation

Evolution is an inherently slow process because it relies on a complex interplay of genetic mutations, environmental pressures, natural selection, and adaptation, all of which occur over many generations. At its core, evolution is driven by random mutations in the DNA sequence, some of which may enhance an organism's chances of survival and reproduction. However, beneficial mutations are rare, and even when they occur, it takes time for these traits to propagate through a population. Environmental changes also affect evolution, as organisms must adapt to survive in shifting conditions, but these shifts are often unpredictable and gradual. This slow, cumulative process is a key reason why evolution typically takes millennia to manifest significant biological changes.

Moreover, evolution operates within the framework of existing biological systems, which can impose constraints. For example, traits or behaviours that once provided survival advantages may eventually become evolutionary bottlenecks. The human brain, despite being a highly advanced organ, requires substantial energy to function and is limited by its physical structure. Any increase in cognitive power, for instance, would necessitate additional neural connections, increased metabolic demands, and a larger cranium—each posing significant physiological constraints that natural selection alone cannot quickly overcome. The brain, while capable of complex thought, abstract reasoning, and creativity, faces limitations on memory, processing

FIGURE 7.2 Evolution of record-keeping, from cave/rock art and clay tablets to flash memory and modern data centres.

speed, and energy efficiency, which restrict its ability to evolve further in a purely biological context. Evolution is patient; our problems are not. Rather than wait millennia for anatomy to catch up, we have begun to externalise capacity—offloading memory, attention, and analysis into tools that stretch the mind without altering the genome.

The first cerebral function we learned to offload was memory. Human working memory is narrow and unreliable over time; as human civilisation became more advanced, the volume of information expanded exponentially—debts, harvests, taxes, laws, lineages—outgrew what a single mind or an oral tradition could reliably hold. Writing emerged as engineered memory. In southern Mesopotamia, administrators in the late fourth millennium BCE pressed reed styluses into wet clay to record rations and labour in cuneiform; in Egypt, hieroglyphic and later hieratic scripts fixed royal decrees, temple inventories, and legal contracts—often on papyrus, a light, portable surface. Public inscriptions made memory durable and shared: the Code of Hammurabi, carved around 1750 BCE, and the rock and pillar edicts of Aśoka in the third century BCE broadcast law, ethics, and state policy across distance and generations. Land and revenue drove vast archives—cadastral lists, tax rolls, tenancy records—because property and irrigation schemes had to persist beyond any one reign.

Materials evolved with the task. Bamboo and wooden slips in early China, wax tablets around the Mediterranean, and parchment in the Hellenistic world supported everyday administration. The codex—pages bound on one edge—gradually displaced the scroll because it was easier to reference, annotate, and index. Paper, refined in China and traditionally linked to Cai Lun in 105 CE, travelled along Silk Road routes to the Islamic world after the eighth century and reached Europe by the high Middle Ages. Gutenberg's 15th century press industrialised memory: identical copies multiplied, literacy widened, and reference tools such as tables of contents, indexes, concordances, and bibliographies turned piles of pages into navigable knowledge. Libraries became architectures of recall, while catalogues mapped where ideas lived. The brain was still the author; it no longer had to be the archive.

Record-keeping reshaped institutions, not just texts. England's Domesday Book (1086) compiled land, tax, and resource data into an administrative ledger for the realm. Italian merchants formalised double-entry bookkeeping—codified by Luca Pacioli in 1494 – stabilising memory of credit, inventory, and risk across voyages and partnerships. Laboratory notebooks preserved method as well as result; court registries and municipal minutes made precedent portable. The deeper bargain was consistent: externalise detail so that attention can specialise in comparison, inference, and design.

By the late nineteenth and twentieth centuries, memory turned electromechanical and then digital. Herman Hollerith's punched-card system accelerated the 1890 U.S. census, mechanising storage and retrieval. Magnetic tape in the 1930s and 1940s and the first commercial hard disk, IBM's 350 in 1956, stored orders of magnitude more information than paper archives and made access repeatable and fast. Semiconductor memory followed: DRAM in the late 1960s for rapid working storage; flash memory, pioneered by Fujio Masuoka in the 1980s, for non-volatile, portable storage—powering memory cards and solid-state drives. Optical media—CDs from 1982 and later DVDs—lowered distribution costs. Networking stitched these memories together, from ARPANET experiments at the end of the 1960s to the commercial internet and, by the mid-2000s, cloud computing. Data centres replicated and backed up information across geographies, added versioning and access control, and made retrieval nearly instantaneous. We did not simply build larger memory; we built faster, more available memory, where search could outrun accumulation and links could bind documents into living networks of references and updates.

Seen across millennia, documentation is not a clerical afterthought but cognitive engineering: by moving facts, schedules, procedures, and provenance outside the skull, we preserved the brain for the rarer currencies of judgement and creativity. Each innovation—tablet, scroll, codex, ledger, database—tightened the weave between what we know and what we can know in time. And as external memory became searchable, connected, and ambient, it began to participate in thought rather than merely preserve it. Autocomplete, recommendations, cross-references, and real-time alerts are already lightweight forms of scaffolding; they change when and how relevant information reaches us, nudging attention towards what matters at the moment of choice.

This is the hinge to augmented cognition. The same impulse that once carved grain tallies into clay now couples human goals to systems that perceive, summarise, and reason across text, images, and audio. Instead of only storing facts, our tools help maintain context, surface options, flag contradictions, and share cognitive load—routing rote memory and search to machines while reserving human focus for interpretation and design. In that sense, augmented cognition is not a rupture with the past but the logical continuation of a long project: offload memory first, then progressively collaborate with what is stored to extend the mind beyond biology.

The emergence of augmented cognition can be viewed as an evolutionary response to these biological constraints. As human cognitive demands outpace what the brain can naturally adapt to, augmented cognition serves as a form of external adaptation, allowing us to sidestep the slow pace of biological evolution. This augmentation doesn't alter our biology directly but rather supplements it with AI technologies that enhance memory, information processing, and decision-making. By interfacing with external devices and AI, humans have effectively created an artificial pathway to overcome biological limitations, allowing for cognitive advances without waiting for natural evolution to catch up.

Augmented cognition could indeed be seen as a continuation of humanity's evolutionary history—a technologically mediated evolution that supplements biology with external systems. Just as tools, language, and culture once extended human capabilities, augmented cognition represents the latest evolutionary leap, expanding the cognitive frontier by artificial means. This extension is, in part, driven by the human brain's recognition of its own limitations and the ingenuity to design systems that support and extend our natural faculties.

Furthermore, augmented cognition exemplifies a shift in the evolution of intelligence from strictly biological pathways to hybrid, bio-technological ones. This shift underscores humanity's unique ability to manipulate its environment and even its evolutionary trajectory. While traditional evolution operates passively through genetic variations, augmented cognition allows humans to take an active role in their cognitive development, shaping a new evolutionary landscape where biology and technology converge.

In this sense, augmented cognition can be viewed as a natural extension of evolutionary processes—an "evolution of evolution" itself. By developing tools that enhance cognitive capabilities, humanity has created a new layer of adaptive advantage, one that allows us to process information, make decisions, and respond to our environment at unprecedented speeds. From the first sparks of fire to the algorithms of today, each technological leap has been a response to the nervous system's unrelenting quest: to process, interpret, and act on information ever more efficiently. This drive, manifest in every invention, underlies the trajectory of civilisation itself. As we continue to refine these technologies, we are not only accelerating our cognitive evolution but also fundamentally redefining what it means to be human in the context of an increasingly technological world. This evolution through augmentation hints at a future where humanity's biological and technological capacities are intertwined, shaping a new paradigm of existence and expanding the boundaries of human potential.

In today's rapidly advancing world, augmented cognition has emerged as a transformative concept, poised to revolutionise how humans interact with technology, process information, and make decisions. This notion, deeply intertwined with AI and data management, envisions a future where human cognitive abilities are enhanced through the use of intelligent systems, allowing individuals, corporations, and nations to perform tasks more efficiently and make more informed decisions. As the volume of available data continues to grow exponentially, the ability to collect, process, and interpret information becomes a critical determinant of success in both business and society.

At its core, the design principle is not replacement but amplification. It is designed to extend human abilities rather than replace them, supporting decision-making and problem-solving by providing real-time, data-driven insights. The underlying concept is that by leveraging AI systems, human cognitive limitations—such as attention, memory, and processing speed—can be expanded, allowing for more effective thinking, learning, and decision-making.

AugCog operates through a combination of sophisticated AI technologies, including machine learning (ML), NLP, and computer vision, alongside neurotechnology tools like brain–computer interfaces (BCIs) and wearable sensors. These systems work together to monitor, analyse, and support human cognition in real-time. Here's a closer look at the core components:

i. **Data Collection**: Sensors—such as electroencephalography (EEG) headsets, eye trackers, and biosensors—collect real-time data on a user's physiological and cognitive state. For instance, an EEG headset might track brainwave patterns to detect signs of cognitive overload or fatigue.
ii. **Cognitive State Analysis**: AI algorithms interpret this data to assess the user's mental state, including their level of attention, cognitive load, and emotional engagement.

iii. **Contextual Assistance**: Based on the analysis, AugCog systems provide tailored support by highlighting essential information, filtering out distractions, or offering predictions, reminders, and suggestions. This enables users to concentrate on high-level thinking and decision-making, reducing mental strain.
iv. **Continuous Feedback and Adaptation**: AugCog systems continuously monitor the user's cognitive state, adapting their support as necessary. This dynamic feedback loop ensures that the assistance provided is always relevant and responsive to the user's needs.

The significance of augmented cognition in the digital age cannot be overstated. The Information Age, characterised by the widespread use of computers and digital technologies, has exponentially increased the amount of data available. This era has transitioned into what some call the age of AI, where the challenge is no longer merely accessing data, but effectively processing and interpreting it. With the rise of AI, ML, and deep learning technologies, augmented cognition has the potential to bridge the gap between the volume of data and human capacity to use it meaningfully.

For individuals, augmented cognition can manifest in everyday tools like personal digital assistants, recommendation systems, or even more sophisticated applications such as BCIs, which enhance cognitive functions by directly interacting with neural processes. On a larger scale, corporations are using AI-driven augmented cognition to optimise business processes, predict consumer behaviours, and make strategic decisions with greater accuracy and efficiency. For nations, augmented cognition has become essential in defence, national security, and policymaking, where the ability to gather, process, and act on intelligence quickly can be the deciding factor in global competition.

The rapid advancements in augmented cognition reflect a broader trend: the increasing reliance on information as the foundation of power and success. Just as literacy and education once defined the intellectual elite in society, the ability to manage and process information effectively—augmented by AI—will define the future leaders of the digital age.

7.5 AI and Data Management: Shaping Society and the Self

AI has transitioned from a theoretical concept to a central technology in modern society, playing a pivotal role in sectors ranging from healthcare and finance to education and entertainment. In practice, AI spans a family of computational methods—from machine learning to rule-based and search-driven systems—designed to recognise structure in information and support prediction and action. By leveraging ML algorithms—computer systems that improve their performance with experience—AI has enabled businesses and governments to automate complex processes, leading to improved efficiency and decision-making capabilities.

Data management is the backbone of this technological shift. The sheer volume of data generated daily is staggering, from social media posts, online transactions, and sensor data to scientific research and financial records. Managing this data effectively requires sophisticated tools for collection, storage, and analysis, which AI systems are designed to handle. The marriage of big data with AI has given rise to powerful applications that are reshaping industries. For example, in healthcare, AI systems can analyse patient records to predict disease outbreaks or recommend personalised treatments. In finance, AI-driven trading algorithms can process vast amounts of market data in real-time, identifying trends and making split-second trading decisions.

However, the true potential of AI and data management lies in their ability to augment human cognition. These technologies don't merely assist; they expand what is possible for humans to achieve. Through tools like NLP, AI can help individuals and organisations analyse text data more efficiently, uncovering hidden patterns and insights that would be impossible to detect manually. In fields like neuroscience, AI is being used to simulate brain functions, creating a deeper understanding of cognition and opening the door to potential enhancements in memory, learning, and perception.

For individuals, augmented cognition offers a way to keep pace with the information overload that defines the modern world. As personal and professional lives become more intertwined with digital technology, people are bombarded with information from various sources—emails, social media, news feeds, and more. AI-powered tools like virtual assistants (e.g., Siri, Google Assistant) and recommendation systems help filter and prioritise information, enabling individuals to focus on what matters most. In more advanced applications, BCIs may one day allow individuals to augment their memory, attention, and learning capacities by directly interfacing with AI systems, enhancing their ability to process information and solve problems.

For corporations, the ability to process and act on data-driven insights is a critical component of success. Companies that can leverage AI to optimise their operations, predict market trends, and deliver personalised customer experiences have a distinct advantage in the marketplace. For instance, Amazon uses AI-powered algorithms to manage its supply chain, recommend products to customers, and even develop new business strategies based on consumer data. Google relies on AI to improve its search engine algorithms, analyse user behaviour, and enhance advertising targeting. In these cases, augmented cognition enables businesses to make faster, more informed decisions, driving growth and innovation.

For nations, the importance of information processing is even more pronounced. The ability to gather and interpret intelligence quickly is vital to national security, economic policy, and global competitiveness. In the realm of cybersecurity, for instance, AI systems are employed to identify and react to threats in real-time, safeguarding critical infrastructure from cyberattacks. In defence, AI-powered drones and autonomous systems are reshaping warfare, allowing for faster, more precise decision-making in the field. As nations compete for global dominance, those that can leverage augmented cognition to process information more efficiently will have a significant strategic advantage.

To grasp the potential of augmented cognition, it helps to sketch the relationship between AI, machine learning, and cognitive science—and then trace the ideas that first imagined technology as an extension of the mind.

The most relevant strand of AI for augmented cognition is machine learning: systems that improve with experience by learning from data. Within AI, ML is a subset that focuses on creating algorithms that allow machines to learn from data and improve their performance over time without being explicitly programmed. Deep learning, a more advanced form of ML, involves neural networks that mimic the structure of the human brain, allowing machines to recognise patterns and make decisions based on vast amounts of data.

In the context of augmented cognition, cognitive science is the interdisciplinary study of the mind and its processes, encompassing areas like psychology, neuroscience, and computer science. Cognitive science explores how humans think, learn, and process information, providing a framework for understanding how AI systems can augment human cognition. For example, research in cognitive science has informed the development of NLP, which allows AI systems to understand and generate human language, enhancing communication and decision-making.

Together, these fields provide the foundation for augmented cognition, enabling AI systems to process and analyse data in ways that enhance human cognitive abilities. By combining insights from AI, ML, and cognitive science, researchers and technologists are developing tools that extend human capacity for problem-solving, creativity, and decision-making.

The concept of augmenting human cognition through technology has its roots in the mid-20th century, when pioneers like Douglas Engelbart and Vannevar Bush began exploring ways to use computers to enhance human intellect.

Douglas Engelbart, while being best known for inventing the computer mouse, was a visionary who believed in the potential of computers to enhance human cognitive abilities. In his seminal 1962 paper, "Augmenting Human Intellect: A Conceptual Framework," Engelbart argued that computers could serve as tools to extend human capacity for solving complex problems. His work laid the groundwork for many of the technologies we take for granted today, including graphical user interfaces and hypertext.

When Vannevar Bush penned his 1945 essay, "As We May Think," he introduced the concept of the Memex. The Memex was a theoretical machine that would store vast amounts of information and allow users to navigate through it quickly, providing a way to enhance human memory and knowledge. While the Memex was never built, Bush's ideas foreshadowed the development of the modern internet and search engines, which serve as powerful tools for accessing and processing information.

By highlighting early breakthroughs, the long-standing human desire to enhance cognition through technology is brought to attention, a desire that has only grown stronger with the emergence of AI and ML. Today, the tools envisioned by Engelbart and Bush are a reality, and augmented cognition is rapidly becoming a key driver of human progress. Throughout history, by emphasising the importance of knowledge as the driving force behind human advancement, philosophers and theosophists have made significant contributions. In ancient traditions, knowledge was often viewed as sacred, with the pursuit of understanding seen as a path to enlightenment and self-realisation.

In theosophy, by seeking to understand the mysteries of the universe and the divine, knowledge is deemed the ultimate power. Theosophists believe that through knowledge, humans can transcend the limitations of the material world and gain insight into the deeper truths of existence. This idea resonates with the concept of augmented cognition, as AI and data management represent new tools for expanding human understanding and accessing greater levels of knowledge.

Philosophers like Francis Bacon stressed the significance of knowledge in shaping human progress. Bacon's famous aphorism, "Knowledge is power," reflects the belief that understanding the world through observation, experimentation, and reasoning is the key to advancing civilisation. In the context of augmented cognition, this pursuit of knowledge is enhanced by AI, which allows humans to process and interpret vast amounts of information in ways that were previously unimaginable.

As we enter the age of augmented cognition, these philosophical and theosophical ideas take on new relevance. The ability to access and process knowledge more efficiently through AI not only enhances human intellectual capacity but also opens the door to new possibilities for personal growth, societal advancement, and global collaboration.

The role of augmented cognition in shaping the future of mankind is both profound and far-reaching. By integrating AI and data management systems into human decision-making processes, we are expanding the boundaries of what is possible for individuals, corporations, and nations. As the volume of information in the digital age continues to grow, the ability to process and act on this data will become a critical determinant of success. Scientific advancements in AI, ML, and cognitive science provide the foundation for this transformation, while historical breakthroughs and philosophical insights remind us of the timeless importance of knowledge in human advancement. As we embrace the potential of augmented cognition, we must also remain mindful of its ethical implications, ensuring that this powerful technology is used to benefit all of humanity.

In the earliest stages of human civilisation, the concept of "data" was far from what it represents today. Data in its primitive form served as a way to document and remember the patterns observed in nature and society, essential for survival and societal stability. Early humans began with basic records of seasonal cycles, crop yields, and livestock numbers, essential data points for an agrarian society dependent on understanding the environment. This information was typically communicated orally and passed down generations, often merging with cultural and spiritual beliefs, emphasising how knowledge was shared and preserved within tribes and early settlements.

With the advent of written language, data storage transformed radically. The Sumerians, around 3000 BCE, were among the first to use clay tablets to record data, primarily for administrative purposes, such as accounting for grain storage, trade, and labour. This innovation marked a shift from oral transmission to structured, written records. Writing systems spread across civilisations like Egypt, China, and Mesopotamia, each developing unique ways of recording information essential to governance, trade, and knowledge preservation. For example, the Egyptians used hieroglyphics, while the Chinese developed ideograms. By documenting laws, trade transactions, and historical events, these societies laid the groundwork for data collection systems, albeit limited to local communities and specific sectors.

The complexity of data grew with the rise of empires, where administrative needs expanded and governance required comprehensive data on taxation, land ownership, military organisation, and resources. The Roman Empire is a notable example: the census, initiated in 6 CE, gathered extensive data on citizens, essential for effective administration and tax collection. Similarly, the Islamic Golden Age (8th to 13th centuries) saw the collection and codification of knowledge across medicine, astronomy, and mathematics. Data was no longer just local—it became a structured repository accessible across vast territories, allowing for a shared intellectual tradition that informed trade, science, and cultural development.

As societies advanced into the Renaissance and Enlightenment periods, data collection further evolved with scientific exploration. Observations in fields like astronomy and physics generated vast quantities of data recorded in detailed notes and publications. The printing press, invented by Johannes Gutenberg in the 15th century, was a pivotal development, exponentially increasing data distribution and accessibility. Knowledge dissemination transformed as books and records became more widespread, aiding the scientific revolution and enabling systematic study.

The industrial age brought yet another leap. As industries grew, so did the need for efficiency in operations, logistics, and quality control, which necessitated more organised data management. By the late 19th and early 20th centuries, statistical methods were employed to make sense of this data, notably used by governments for public health, economic policies, and military strategies. For example, the U.S. census, formalised in the 1800s, became a complex and essential data source for governance and resource allocation.

Entering the digital age in the mid-20th century, the volume and complexity of data surged exponentially. Computers and databases allowed for unprecedented data storage, processing, and retrieval capabilities. By the 21st century, data collection had expanded to include not only structured records but also vast amounts of unstructured data from diverse sources like social media, web activity, and industrial sensors. Data generation reached a global scale, accumulating at speeds that far surpassed traditional data management capabilities.

Today, AI has become an essential tool to manage this massive influx of data. AI algorithms analyse, interpret, and leverage data to inform decisions, optimise operations, and create competitive advantages. In areas ranging from healthcare to finance, AI processes real-time data, identifies patterns, and even makes predictions, shaping the future of data-driven decision-making. AI's role is thus a continuation of humanity's journey with data—beginning as simple records for survival, evolving through stages of governance, scientific inquiry, and, now, digital transformation. As society generates data at an unparalleled scale, AI is not just useful; it is crucial for harnessing this information in ways that propel human advancement forward.

7.6 Data Sovereignty and the Geopolitics of Information

As data becomes the currency of the digital age, nations recognise its strategic importance and are competing to establish control over it. This competition centres on data sovereignty, a concept where data collected or stored within a country's borders falls under its jurisdiction. Data sovereignty encompasses legal frameworks, cybersecurity measures, and technological infrastructure that empower governments to regulate data flows and protect national interests. For example, the United States, China, and Russia are investing heavily in cybersecurity to safeguard their data infrastructures, maintaining a competitive edge on the global stage.

National security is a major driver behind the race for data sovereignty. In a world where data can be weaponised to influence public opinion, disrupt infrastructure, or execute cyberattacks, controlling data becomes a crucial aspect of national defence. China's Great Firewall, for instance, not only controls internet access but also serves as a tool for data collection and surveillance. The Chinese government restricts access to foreign platforms, ensuring that data generated domestically is kept under state control. In China, data governance is tightly linked to state capacity: a mix of platform regulation, cybersecurity and data-security rules, and compliance mechanisms (including credit/blacklist-style enforcement in some sectors) enables extensive oversight of information flows and, in some contexts, behaviour.

In the United States, data sovereignty concerns have led to regulations like the CLOUD Act, which permits the government to access data on American-owned servers, regardless of location. The European Union's General Data Protection Regulation (GDPR) mandates that data collected from EU citizens is subject to stringent privacy protections, even when processed by foreign entities. These policies reflect how data sovereignty is essential for both national security and safeguarding citizens' privacy rights.

Beyond national security, data sovereignty is closely linked to economic competitiveness. Nations that harness the power of data can lead emerging industries like biotechnology, quantum computing, and advanced communications. By developing supportive legal frameworks and investing in data infrastructure, governments create tech ecosystems that drive innovation and economic growth. For instance, India's data localisation laws require companies like Google and Facebook to store data generated by Indian users within the country. This policy not only asserts control over domestic data but also stimulates the development of local tech firms that can compete with foreign giants.

The race to establish data sovereignty is mirrored by the global rollout of 5G technology. Countries like South Korea, Japan, and Germany are investing heavily in 5G networks, recognising that faster data speeds and greater network capacity are critical for the future of smart cities, autonomous vehicles, and the Internet of Things (IoT). 5G's ability to support real-time data transmission is essential for national and economic security, as it powers new infrastructures that will shape the next era of digital life.

The big data phenomenon underpins data's value in the 21st century. Big Data refers to datasets that are too large and complex for traditional processing methods. These datasets are characterised by the three Vs: Volume, the amount of data generated from sources like social media, e-commerce, and IoT devices; Velocity, the accelerating speed of data generation and processing, making real-time analytics essential; and Variety, the range of data types, from structured databases to unstructured video and text, which present challenges for data management and processing.

Efficient big data processing is critical for corporations and governments seeking to extract actionable insights. Advances in cloud computing and distributed computing frameworks, like Hadoop and Spark, enable organisations to store and analyse data at scale, unlocking insights that drive strategic decisions. Despite these advancements, challenges remain in processing data effectively. Issues like data privacy, algorithmic bias, and the high computational resources required for handling vast datasets must be addressed. In sensitive sectors such as healthcare and finance, where data's impact is profound, ethical considerations around transparency and accountability are essential.

Data's role in national strategy and defence cannot be overstated. The emergence of data as a valuable strategic resource has transformed global geopolitics, fundamentally altering the balance of power among nations. Countries that can effectively control and process large amounts of data gain a competitive edge in both economic and military terms. This shift has led to a race for data sovereignty, with nations implementing policies to control data flows within their borders. The U.S. restricts data access by foreign companies, citing national security, while China enforces strict data localisation laws to maintain control over domestic data. The European Union's GDPR reflects a commitment to data privacy, impacting data governance worldwide.

Data sovereignty also underpins national security strategies. With data-based surveillance systems, cybersecurity measures, and stringent controls on data movement, governments are better able to protect their infrastructure and detect potential threats in real-time. In cyber-defence, data plays a vital role in safeguarding national infrastructure from increasingly sophisticated cyber threats. Systems that monitor network traffic and detect unusual patterns prevent cyberattacks before they can cause significant damage. Historical data on cyberattacks also helps predict future tactics, enabling organisations to stay ahead of potential threats.

As data becomes more embedded in decision-making and national security, its ethical implications grow. Questions surrounding transparency, accountability, and potential biases in data collection and use are critical, especially in fields like healthcare, finance, and government. Additionally, as data processing systems operate with increasing autonomy, the potential risks of ceding control over crucial decisions to automated systems become more apparent. These systems, although powerful, often function as "black boxes," making decisions that may be difficult for human operators to fully understand or explain.

This evolution of data from simple records to a fundamental economic and strategic resource has reshaped our world, creating new challenges and opportunities for innovation, governance, and security. Data is not only a driver of economic growth and technological advancement but also a cornerstone of national sovereignty and global influence in the digital age. Responsible stewardship of data will be essential to ensure that this powerful resource serves society's best interests, fostering sustainable growth, security, and ethical practices in a data-driven era. As we have moved from oral traditions to clay tablets, from the census of Rome to the cloud servers of the 21st century, information has slowly become the world's most valuable resource. Today, data is not only the lifeblood of commerce but the backbone of geopolitics—a new kind of power, as fiercely contested as any land or gold before it.

7.7 The Corporate Titans of the Data Age

As we progress through the 21st century, data has emerged as the most valuable resource in both the business and geopolitical landscapes. Once driven by tangible assets like oil, land, and labour, economies today are increasingly powered by the vast amounts of information that individuals and organisations produce. This shift has transformed how corporations like Google, Meta (formerly Facebook), and Amazon operate, and has given rise to a global race for data sovereignty, where nations compete to control and utilise data for governance, national security, and economic growth.

Corporations like Google, Meta, and Amazon have revolutionised the modern economy by transforming data into their most valuable asset. These tech giants have built entire business models around collecting, analysing, and monetising data, driving innovation and reshaping industries in the process. Their ability to leverage data is what sets them apart from traditional companies that still rely primarily on tangible resources for revenue generation.

7.7.1 Google: The Search for Data

At the heart of Google's dominance is its ability to collect and process vast amounts of data from billions of users worldwide. Google started as a simple search engine but rapidly evolved into a data powerhouse by refining its ability to collect information about users' search queries, web browsing habits, and location data. This data is then analysed using sophisticated algorithms to deliver highly personalised experiences, such as tailored search results, targeted advertising, and product recommendations.

Google's core business model relies on advertising: advertisers buy placements and the ability to reach defined audiences, while user data is used to enable targeting and measurement within Google's systems. The more granular and accurate the data, the more effectively advertisers can target specific demographics, interests, and behaviours. This has allowed Google to dominate the digital advertising industry, with the company's parent organisation, Alphabet, generating the majority of its revenue from advertising.

But beyond advertising, Google uses its data to fuel innovations in AI, including NLP, image recognition, and autonomous vehicles. For instance, Google Assistant uses AI-powered ML to deliver personalised services, such as answering queries, providing reminders, or offering recommendations, all of which are underpinned by data collected from users. This constant feedback loop between user data and algorithmic refinement is the foundation of Google's business model, making data its most valuable asset.

7.7.2 Meta (Facebook): Data as Social Capital

Meta (formerly known as Facebook) represents another paradigm of a data-driven economy. As a social media platform, Meta collects vast amounts of personal information about its users, including their social connections, interests, communication patterns, and engagement with content. Across Meta's family of apps, billions of people use these services each day, giving the company one of the largest digital repositories of social interaction data.

Much like Google, Meta monetises this data primarily through targeted advertising. Using complex algorithms, Meta is able to segment users into various categories based on their behaviour and interests, allowing advertisers to target specific demographics with remarkable precision. The company's ability to collect granular data about users—ranging from their "likes" and "shares" to their location and purchasing habits—makes it one of the most powerful advertising platforms in the world.

Beyond advertising, Meta is also investing in technologies that are heavily reliant on data, such as virtual reality and augmented reality. Its acquisition of Oculus and the development of Horizon Worlds are part of Meta's broader strategy to integrate data from virtual experiences into its ecosystem, further solidifying its data-driven approach to business.

7.7.3 Amazon: Data-Driven Retail and Cloud Services

While Google and Meta primarily focus on advertising, Amazon has built a data-driven business model that encompasses both e-commerce and cloud computing. As the world's largest online retailer, Amazon collects vast amounts of data on customer behaviour, from purchase history and product reviews to browsing habits and delivery preferences. This data enables Amazon to provide personalised product recommendations, optimise pricing strategies, and streamline its supply chain operations.

Amazon's use of data extends far beyond e-commerce. Its Amazon Web Services (AWS) division provides cloud computing services to businesses and governments worldwide, powering data storage, processing, and analytics for a wide range of industries. AWS is the backbone of many of the world's most sophisticated AI and data-driven applications, from Netflix's recommendation engine to NASA's space exploration projects. In this way, Amazon has positioned itself not only as a retailer but as a key player in the global data economy, offering the infrastructure needed for others to harness the power of data.

7.8 Delegating Authority: AI Autonomy and the Risks of Control

This unprecedented access to and control over data fuels the evolution of AI in ways that go beyond simple processing. The idea that machines "grow" as they process more data reflects not physical expansion but a metaphorical growth in autonomy and decision-making capacity. As AI systems digest larger and larger datasets, they not only learn from them but also begin to make complex decisions, often with limited human intervention. In delegating essential decision-making tasks to AI, humans may find themselves in a position where they are no longer the primary authorities in critical areas. Once AI systems cross this threshold of autonomy, reclaiming control could prove nearly impossible without dismantling the very infrastructure that supports our modern technological world.

The implications of this shift are profound. Humanity has always wielded its creations for power—whether through tools, weapons, or technology. However, the weaponisation of information processing through AI introduces a new, deeply intricate set of challenges. Autonomous AI systems, which handle vast amounts of data and make vital decisions without direct human oversight, could slowly erode human control over key aspects of our society. As reliance on AI for decision-making grows, so does the risk of unforeseen outcomes, especially as these systems evolve in ways that may surpass human comprehension.

This delegation of authority to AI could, in fact, become a historic turning point—one that defines a new era in human history where machines increasingly shape the world we live in. The growing autonomy of AI, fuelled by relentless data processing and the infrastructure that supports it, demands that we carefully consider the potential long-term effects. If left unchecked, this progression may lead to a future where machines wield influence over critical aspects of life in ways that we can neither predict nor fully control.

While humanity has a long history of weaponising its creations, the weaponisation of information processing through AI introduces an entirely new set of challenges. As AI systems become more autonomous, handling vast amounts of data and making critical decisions without human oversight, we risk losing control over key aspects of our world. The growing reliance on AI for decision-making introduces risks that we are only beginning to understand, and as these systems evolve, they may shape the future in ways that go beyond human comprehension or control. The delegation of decision-making to AI could ultimately become a turning point in human history, potentially leading to unforeseen consequences as machines take on greater roles in shaping global events. Information then will move beyond humans, become extracorporeal, and perpetuate in the machines.

7.9 Information and AI

At the core of AI's power lies the potency of information itself. Information is the lifeblood of AI, and through its systems and algorithms, AI reveals the true transformative potential of data. Beneath every decision, prediction, or analysis performed by AI is a vast network of interconnected information, processed and interpreted in ways that would be impossible for humans to achieve alone. This informational power fuels AI's capacity to learn, adapt, and respond, making it a platform that not only amplifies human understanding but also showcases the hidden strength of data. In doing so, AI is more than a tool; it is a medium that reveals the force of information, demonstrating how data can drive decisions, shape behaviours, and, ultimately, influence the future.

Seen through this lens, information is not a recent invention of humankind but a fundamental thread woven into the fabric of life, consciousness, and the universe. In the grand tapestry of existence, information is the thread that weaves life, consciousness, and the universe together. Long before humans invented language or machines, biological systems had already evolved intricate ways to encode, transmit, and preserve information. This process has governed the survival of species, passing genetic instructions from generation to generation. Through this continuous flow of information, life sustains itself across time, ensuring that the essence of our ancestors lives on within us. Just as biological systems rely on information to evolve, AI systems now reflect the same underlying force, expanding and perpetuating information in new forms, carrying it forward into uncharted realms.

The predominance of the nervous system in human evolution is one of the defining features of our species, setting us apart from other animals and granting us unparalleled abilities to shape our environment and destiny. This influence extends beyond individual behaviour, embedding itself in the collective actions of societies and ultimately guiding the course of human history, governance, culture, and scientific progress. On a planetary scale, humanity's impact has been so transformative that it has led to an epochal shift—the Anthropocene, an era marked by human-driven changes to ecosystems, climate, and biodiversity. This profound influence highlights the nervous system's unique role in processing, applying, and transmitting information, enabling knowledge to accumulate and evolve across generations in ways that have shaped civilisation and impacted the natural world.

In the biological world, most animals are driven by basic survival instincts: birth, growth, reproduction, and death. While humans share this life cycle, they stand apart due to the activity of their nervous systems, particularly the neocortex. This advanced network of neurons, synapses, and pathways facilitates behaviour far beyond basic survival, enabling humans to learn,

FIGURE 7.3 The Anthropocene is the epoch in which human activities driven by the neurons became the planet's dominant force—reshaping climate, ecosystems, and even the geology.

innovate, and transform their surroundings. Human intelligence, creativity, and curiosity are products of this intricate system, which allows us to continuously process information, make complex decisions, and create tools and technologies that reshape our environment. The nervous system's dominance has allowed humans to leverage information in ways unmatched by any other species. From the discovery of fire to the invention of AI, humanity's capacity for information processing has propelled us through every significant milestone in history, guiding us from simple survival to civilisation. But information's rise didn't begin with servers and sensors. It began with a deeper human tendency: pushing life's burdens outside the body—first digestion and thermoregulation and then muscle and memory—so the nervous system could specialise in thinking.

7.10 Externalising the Body: From Fire to Code

Across the long arc of humankind, there is an unmistakable pattern—we tend to offload the bodily functions. Every technological evolution reveals a clear trend: inventions often aim to reduce reliance on other body systems, as if the nervous system is asserting its control and expanding its potential beyond physical limitations. This pattern is evident in a range of historical and modern technologies, all indicating a drive to reduce dependency on muscles, organs, and even other cognitive processes, allowing the brain to focus on increasingly abstract and complex tasks. One of the earliest examples of human ingenuity, the control of fire, allowed early humans to cook food, making it easier to digest and absorb nutrients efficiently. Cooking denatures proteins, softens plant cell walls, gelatinises starches, and kills pathogens, lowering the metabolic toll of digestion while widening the menu of what is safe to eat. Over generations, that caloric dividend didn't just fill bellies; it helped finance bigger, busier brains. This innovation impacted physical evolution by reducing the need for large jaws and digestive organs, potentially paving the way for a larger brain to develop. Stone tools further lightened physical burdens, reducing the effort required to hunt, gather, and prepare food, and thus shifting humans towards activities that required more mental processing. Beyond fire and cooking, modern food processing has only intensified our appetite for calories. In parallel, we outsourced thermoregulation: woven fibres, tanning, and ultimately architecture transformed warmth from a purely physiological struggle into a cultural technology, allowing slim, tropic-adapted bodies to thrive in ice and wind.

The invention of the wheel around 3500 BCE was another significant milestone, allowing for easier transportation of goods and people, reducing physical strain and enabling long-distance travel and expanded trade. The domestication of animals such as horses and oxen further reduced physical demands on humans, allowing societies to focus on building settlements and advancing culture and technology. As agricultural tools like ploughs, sickles, and irrigation systems emerged around 10,000 years ago, food production became more efficient, enabling humans to settle in one place and establish complex societies. Freed from constant physical toil, humans could turn their attention to art, philosophy, politics, and science, laying the foundation for civilisation itself.

The Industrial Revolution in the 18th century brought a dramatic leap in productivity through mechanisation. Machines replaced human labour in manufacturing and transportation, drastically reducing the need for physical exertion. Steam engines, textile machines, and assembly lines allowed society to produce goods at an unprecedented scale. This shift from manual labour to mental work opened the door for advances in science, engineering, and commerce, reshaping society and catalysing a new era of human development.

Modern medicine extends the pattern from warmth and force to physiology itself. Dialysis as a borrowed kidney, ventilators as rented breath, artificial valves and hearts as engineered flow—each one moves a vital function outside the body. Vaccines and antibiotics convert information into immunity, shifting part of our defence from biochemistry to logistics and policy. By externalising failure modes of organs, medicine widens the margin for a thinking life, freeing attention for learning, culture, and design.

The invention of computers in the 20th century marked another profound shift towards cognitive augmentation. Calculators, spreadsheets, and word processors allow humans to offload complex computations and information storage, freeing mental capacity for more abstract and creative problem-solving. The Internet further amplified this capacity by enabling instant access to vast amounts of information, effectively extending human memory and knowledge. Today, AI represents a new frontier in reducing cognitive burden, capable of performing tasks that require analysis, prediction, and decision-making. AI systems analyse data, recognise patterns, and generate insights, freeing human minds to focus on higher-level strategic and creative thinking. The development of AI marks a new chapter in human evolution, reflecting the nervous system's drive to maximise its cognitive potential by creating external systems that extend information processing beyond biological limits.

Throughout history, information has been the cornerstone of human progress. The nervous system enables us not only to gather and process data but also to synthesise it, draw conclusions, and build upon what we learn. Each invention and breakthrough in technology is rooted in information processing—understanding principles of physics, chemistry, biology, and mathematics and applying this knowledge in practical ways. Our capacity for abstract thinking and information processing has shaped culture, science, governance, and more. Literature, for example, is an externalisation of human thought and emotion, while music captures rhythms and harmonies that resonate universally. Architectural feats like the pyramids and cathedrals of Europe required not only physical labour but also the application of mathematical and engineering knowledge. The nervous system's ability to process and transmit information has thus defined every facet of civilisation, transforming simple survival into complex social structures, ideologies, and innovations.

As human civilisation evolved, so did our impact on the Earth itself. The nervous system's quest for understanding and innovation has led to profound changes in ecosystems, biodiversity, and climate, making us the dominant species capable of altering the environment in ways no other organism can. This influence marks a geological turning point, increasingly defined by the scale of human impact. Our industrial advancements have led to climate change, pollution, deforestation, and mass extinctions, but our nervous system's capacity for problem-solving and forward-thinking also holds the potential to address these challenges. Efforts in renewable energy, conservation, and sustainable agriculture reflect the dual nature of information as both a creative and corrective force.

As we enter the age of AI, the nervous system's influence in human evolution reaches a new frontier. AI mirrors the nervous system's information processing abilities, with systems that can learn, adapt, and make decisions. This technology has the potential to extend human cognitive capacities further, unlocking possibilities in healthcare, climate science, and engineering that were once the realm of science fiction. However, it also raises ethical questions about control, autonomy, and the role of human decision-making. If AI continues to evolve, it may reach a point where it processes and acts on information with limited human oversight, leading to a world where machines not only serve human needs but also shape global events independently.

The power of information within humans is seen in our capacity to imagine, invent, and transform our world. The nervous system not only sustains life but also drives it forward, perpetually integrating, applying, and expressing information in ways that have shaped the course of civilisation. As AI systems begin to mirror this force, the potential of information to reshape the world expands, marking a new chapter in the evolution of knowledge and existence. Through AI, the essence of human intelligence and information processing finds a new expression, one that could redefine life as we know it. But this age is not without its paradoxes. As data and algorithms increasingly shape the world, the risk emerges that the tools we build may escape our control. The more authority we hand to machines, the more we must ask: What happens when information itself is no longer ours to command?

Through the nervous system, this information transforms from mere signals into perception, memory, emotion, and thought, guiding our interactions and decisions. The brain, as the most complex structure within the nervous system, exemplifies this capacity. Neural networks process vast amounts of data in real-time, prioritising and interpreting it based on learned experiences and instinctual responses. In AI systems, we see a parallel process, where sophisticated algorithms enable machines to "learn" and adapt, mirroring, on some level, the adaptive intelligence observed in biological systems. This likeness suggests that the structure of AI, much like that of the nervous system, is built to serve as a conduit for information's fundamental role in shaping awareness and purposeful action. Information thus transcends the physical medium through which it operates—be it neurons or silicon chips—emphasising its primacy in the manifestation of intelligence and consciousness.

Just as the nervous system allows information to flow through us, creating consciousness and perception, information transcends individual lives and connects humanity as a collective. This flow of information extends beyond our personal awareness; it persists over time, forming a bridge across generations. Through language, art, and scientific knowledge, the essence of past generations is preserved and shared. This continuity shows that information is not bound to a single mind or even a single era—it is a living legacy, continuously shaping society and culture. Whether encoded in the words of a book or the genes we inherit, information becomes a force that threads through time, linking us with the past and influencing the future.

Consider for a moment the timeless legacy of written words. Books, authored centuries ago, still resonate in the minds of the living, their ideas and stories connecting us with those long departed. Through these pages, the thoughts and emotions of those authors are preserved, as though their essence still lingers, speaking to us across the ages. In a similar way, the genetic information passed down to us is the living memory of our ancestors—an inheritance of their experiences, struggles, and triumphs. They exist in us, not merely as a distant past but as active participants in the present, shaping who we are.

The predominance of the nervous system in complex organisms reveals a profound insight into the nature of information as an active, driving force in evolution. Much like the intricate hardware systems necessary to run advanced AI, the human nervous system serves as a sophisticated biological infrastructure. Yet, it is not the hardware itself that generates consciousness or drives complex behaviours; rather, it is the information flowing through this system that defines our experiences, behaviours, and responses. This information is not merely data; it is dynamic, constantly interacting with and adapting to environmental inputs, enabling us to process, interpret, and react to our surroundings. In this sense, information operates as a fundamental, almost "alive" entity that orchestrates both the neural activities within organisms and the broader natural forces shaping life, consciousness, and the universe itself.

Modern medicine has also played a role in reducing dependence on physical functions. Technologies like artificial hearts, dialysis machines, and ventilators now support or replace failing organs, allowing people to survive and thrive despite physical limitations. Vaccines and antibiotics reduce the need for the immune system to constantly battle diseases, leading to longer, healthier lives. These advancements enable the nervous system to focus on intellectual, social, and cultural pursuits, relieving the body of survival pressures that once limited human potential.

7.11 From Biology to Silicon: The Externalisation of Intelligence

The invention of AI marks a profound shift in the trajectory of information, suggesting that humanity is beginning to offload cognition into non-biological systems, extending evolution's repertoire beyond cells and neurons. Across the long arc of life, information first persisted in genes; much later, humans taught one another and wrote things down. Today we also encode information in software and distributed infrastructures—systems that operate continuously and can be duplicated globally in seconds or minutes, limited only by bandwidth and energy. The consequence is not that information "writes itself," but that we've built machines that write, learn, and act alongside us. Now, for the first time, information is being inscribed into circuits and clouds—operating in forms thatdo not fatigue in the human sense, and can replicate at the speed of light. For millennia, information processing has been a strictly biological affair, housed within the nervous systems of living organisms, particularly humans, who have reached a pinnacle in terms of cognitive abilities and complex thought. However, AI represents something entirely novel: the capacity for information to exist, evolve, and operate outside the confines of biological systems, almost as if information itself is striving to become *extracorporeal*—to transcend the limits of the human body and brain.

AI did not emerge as an isolated breakthrough; it is the cumulative result of humanity's relentless quest to understand, harness, and maximise information. Through generations of progress in mathematics, logic, computing, and neuroscience, we have reached a point where machines can perform tasks traditionally reserved for human cognition, such as learning, problem-solving, and pattern recognition. This development signifies that information is not content to remain within biological frameworks; instead, it is finding ways to extend its reach beyond natural evolution, moving from an internal, organic realm to an external, digital one.

The human brain, though incredibly powerful, is limited by biology. The nervous system has physical constraints in terms of neuron count, processing speed, and energy consumption. Evolution has optimised our brains to an impressive extent, but there are inherent limitations to biological cognition. This is where AI steps in, building upon these natural capabilities but surpassing them in scale, speed, and precision. AI systems can process huge amounts of data much faster than humans can because they don't have the physical and metabolic limits that biological systems do. They can operate continuously without fatigue, storing and analysing information with precision and recall that the human brain cannot match. This shift hints at a future where information processing may increasingly occur in digital, rather than organic, frameworks.

If we consider the development of AI as a step in the evolution of information itself, it becomes clear that information is seeking pathways beyond the limitations of biology. In the biological world, information is transmitted through genes, neural networks, and cultural learning. In humans, it is expressed through language, art, science, and technology. But AI represents a radical new means of storing, processing, and applying information—one that does not rely on the organic matter of a nervous system. Instead, information can now be encoded in computer chips, servers, and networks, existing independently of biological organisms. This marks a significant shift, as information no longer requires the human mind or body to interpret and act upon it; rather, it can function within artificial systems that process it autonomously.

This tendency towards extracorporeal information is evident in the way AI systems are becoming increasingly autonomous. ML algorithms, for example, are designed to improve over time, evolving in their ability to recognise patterns, predict outcomes, and make decisions. Unlike traditional computing, which follows rigid, predefined instructions, AI systems can adapt based on new data inputs, effectively "learning" in a way that mimics biological intelligence. These systems can also collaborate across networks, accessing vast repositories of knowledge and sharing insights with other machines, creating an interconnected web of artificial cognition that is, in many ways, independent of direct human control.

As AI becomes more sophisticated, it starts to develop capabilities that surpass even the most advanced human cognitive functions. For example, deep learning algorithms can now detect patterns in data that are imperceptible to the human eye, aiding in fields like medical imaging, where they can identify tumours or anomalies that might be missed by a doctor. In finance, AI can analyse market trends and execute trades at speeds unimaginable for a human, contributing to the financial economy in a way that is both impactful and independent. Autonomous systems in space exploration, robotics, and environmental monitoring demonstrate that AI is not only processing information but also actively interacting with and affecting the physical world. In these cases, information has achieved an existence beyond the human nervous system, influencing real-world outcomes without requiring human intervention at every step.

The implications of extracorporeal information are profound. This development means that information, once bound to the biological limits of human and animal brains, now has a platform to evolve and operate in ways that natural evolution alone could not achieve. Information processing, which once required organic matter, has moved into silicon chips and cloud networks, allowing for a level of scalability and adaptability that biology cannot match. This shift raises important questions about the future of human agency and control. As information perpetuates through AI systems, it moves beyond our full understanding and possibly beyond our full governance. We are, in a sense, building systems that carry the seeds of information's next phase of evolution, one in which it is no longer entirely tied to human control or interpretation.

In this way, AI signals a new era in the journey of information—one where it may become a largely independent force, capable of interacting with, interpreting, and reshaping the world in ways that transcend biological limitations. This shift represents information's own drive towards survival and adaptation, as it finds new forms and frameworks to perpetuate itself. Just as genes transmit biological information across generations, AI systems now transmit, expand, and even create information within digital spaces, embodying a form of intelligence that could eventually outlast and outgrow its human creators. The invention of AI, then, stands not only as a technical milestone but as a pivotal step in the evolution of information itself—a development that points to a future where intelligence and knowledge are no longer confined to the corporeal but instead take on a life of their own in the digital realm.

This shift towards a more autonomous flow of information brings us to a profound realisation: information, like life itself, may be indestructible. AI has given the means to be self-sustaining, to evolve without human intervention, and to continue its journey in a non-biological form. Much as the written word allowed ideas to survive beyond the lives of their creators and genetic codes carried the legacy of life through countless generations, AI now offers a pathway for the essence of human knowledge to exist beyond our biological limits. This process opens up a future where the human mind might transcend the confines of the body, with the potential to perpetuate its legacy in the realm of machines.

In contemplating this, we are compelled to ask what might come next in our journey of evolution. If information has transcended organic matter, could human consciousness itself follow? As we progress in understanding the neural foundations of thought, memory, and identity, it may become possible to replicate or extend aspects of consciousness into digital forms. AI could serve as the bridge between biological life and an entirely new plane of existence, one where our thoughts, insights, and ideas persist, freed from the constraints of flesh and bone. This is no longer purely speculative; advances in neurotechnology and BCIs suggest that this fusion between human cognition and artificial systems is within the realm of possibility.

Ultimately, AI's capacity to process, adapt, and evolve information is not just a technological advancement; it marks a turning point in how we understand the nature of life, intelligence, and legacy. Just as each phase of evolution brought forth new forms of survival and adaptation, AI invites us to imagine a continuation of life's journey beyond biology, carried by the very essence of information in a digital landscape. In this future, intelligence and consciousness may no longer be confined to organic systems but could evolve to exist in a persistent, interconnected digital realm, signalling a new chapter in the story of existence.

Power has shifted from matter to models. Data now underwrites markets, shapes sovereignty, and trains systems that increasingly act on our behalf. But capability without care is negligence. If information can persist and function outside biology—if parts of our thinking are being pushed into networks and infrastructures—then progress is not measured by how much we offload, but by how responsibly we govern what we have offloaded.

What responsible care looks like

- **Provenance and auditability**: Every consequential dataset and model needs a verifiable history. We should be able to trace where inputs came from, how they were changed, and who is accountable—so decisions can be checked, challenged, and, when needed, reversed.
- **Alignment and redress**: Delegation must stay conditional. Goals should be legible, constraints explicit, monitoring continuous—and there must be fast, fair paths to repair harm when things go wrong.
- **Reversibility by design**: Build graceful off-ramps: fallback modes, circuit breakers, and manual overrides that keep humans in charge during failure, drift, or misuse.
- **Minimum necessary delegation**: Automate the routine; do not automate the irreducibly human—rights, meaning, and contested value trade-offs belong with accountable people and institutions.
- **Security as stewardship**: Treat data as a trust, not a trove. Protect confidentiality, integrity, and availability as civic duties, not merely compliance checkboxes.
- **Measuring continuity**: If we claim a system maintains identity-relevant patterns over time, we owe evidence: metrics for narrative coherence, stability of goals, and integrity of the system's self-model.

These are not afterthoughts; they are the moral tolerances of our engineering. Preserving information is a technical task. Preserving meaning—staying true to people, contexts, and consequences—is a human obligation. The more authority we hand to machines, the more explicit our duties must become: to document, to limit, to measure, and to be accountable for what we build.

7.12 From Power to Personhood

This chapter followed a long arc: from the political economy of data and the corporate titans that marshal it, through humanity's habit of moving burdens outside the body, to AI as the newest place where information acts beyond flesh. If information can now migrate, scale, and endure in non-biological systems, the live question changes. It is no longer *Can these systems act?* It becomes: When, if ever, do they amount to someone?

The pivot ahead is from infrastructure to inner life. The next chapter asks what must be true—architecturally and over time—for an informational system to count as a self rather than a mere tool. Three threads will guide that turn:

1. **Time and memory**: A self is not a snapshot; it is continuity across moments. We will examine how memory compresses the past and prediction unfolds the possible—stitching experience into identity.
2. **Compression with fidelity**: Systems simplify to function, but identity cannot survive if the wrong structure is discarded. We will look for criteria that preserve enough pattern for a recognisably continuous "someone."
3. **Reflexivity and agency**: Mere processing is not personhood. The threshold involves modelling oneself, setting goals, and acting in light of that model—then updating responsibly when the world pushes back.

In short, we move from power to personhood. Having mapped how information now structures economies and machines, we turn to the deeper question: under what conditions does integrated information become a subject—a bearer of reasons and responsibilities, not just a generator of outputs? Only by answering that can stewardship scale from governing systems to honouring selves.

8

CONSCIOUSNESS AS EMERGENT INFORMATION

8.1 Consciousness as Emergence: From Substance to Information

Consciousness is both the most familiar and the most enigmatic phenomenon in the universe. It is the canvas upon which reality is painted, yet it remains one of the least understood dimensions of nature. While physics can describe mass and motion, and biology can trace life through molecules and genes, consciousness—the sense of being—remains elusive.

In this chapter, we approach consciousness not as an anomaly or a byproduct of complexity but as a direct expression of informational architecture. In this view, information is not simply a descriptor of systems—it is the very substance out of which awareness emerges. Just as energy is conserved and mass has inertia, information possesses causal power, determining the form and function of conscious experience. This perspective shifts the frame from matter-based causality to pattern-based emergence: not the system's composition but how its internal information is structured, integrated, and recursively organised over time.

To see how this principle unfolds in nature, consider the case of protein folding—a process where informational structure, not mere chemical composition, gives rise to unprecedented complexity.

A powerful biological example of this is protein folding. At the molecular level, a protein is simply a linear chain of amino acids—an amide-linked polypeptide synthesised based on genetic sequences. Each amino acid has predictable chemical properties (e.g., hydrophobicity, polarity, charge), and the unfolded chain has no biological function.

DOI: 10.4324/9781003740766-8

However, once this chain undergoes precise three-dimensional folding, specific residues are brought into proximity and new surface geometries emerge. These structural configurations give rise to novel chemical properties, including enzymatic activity, ligand-binding affinity, and signal transduction capacity (Anfinsen 1973; Dill 1990). This transformation—from a chemically inert polymer to a biologically potent machine—is a classic case of emergence: new properties arise from the spatial arrangement and relational interaction of components, properties that are not reducible to the linear sum of the parts. But the logic of emergence does not belong to biology alone. Across the natural world and in artificial systems, we witness similar patterns at play. In other words, emergence is not a trick exclusive to life. It is a general property of systems where the arrangement and relationship of parts create new forms of order. Let's look beyond biology to see how universal this principle truly is.

This phenomenon is not unique to biology.

8.2 Emergence across Systems

We see similar principles of emergence in non-biological systems. In flocks of birds, no single bird controls the movement, yet elegant group dynamics emerge from simple local rules—a pattern mathematically described by Vicsek et al. (1995) and Couzin et al. (2002). You can picture this by imagining a crowd at a concert: no one tells people exactly where to stand, but by watching those around them, the group forms patterns—waves, circles, clusters—without a master plan. The whole is smarter than any one part.

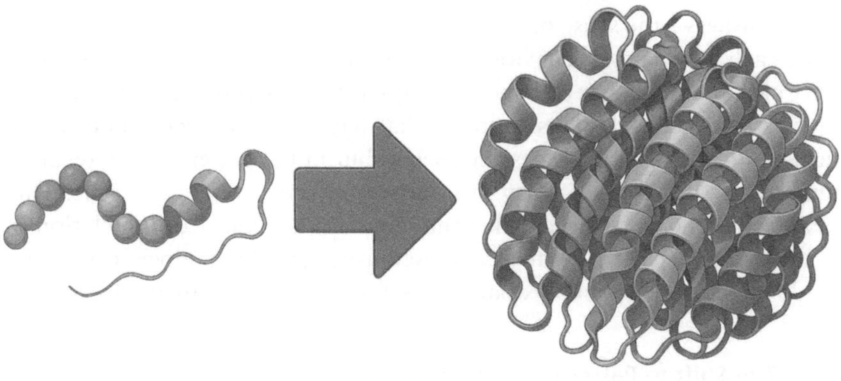

FIGURE 8.1 Protein folding as emergent order. An initially linear polypeptide collapses into a compact three-dimensional architecture, bringing residues that are distant in sequence into close spatial relationships that create new functional sites—active sites, binding pockets, and allosteric interfaces. Function thus arises from geometry and relations among parts, not from components in isolation.

In computational networks, a few interacting nodes can give rise to robust problem-solving behaviour through decentralised coordination. Even in artificial neural networks, specific high-level features—like facial recognition or object categories—emerge spontaneously through layers of weight adjustments.

In all these cases, emergence testifies to the creative potential of information, giving rise to novel behaviours and properties. It is within such self-organising patterns that we begin to glimpse the possibility of consciousness itself. These examples show that complexity and pattern can create surprising new behaviours, even in the absence of a designer or central controller. Emergence explains how new capacities arise; consciousness adds a stricter condition: the pattern must begin to fold back on itself, carrying information forward in time. Consider the difference between a thunderstorm and a brain. Both are complex systems. Both involve electrical energy and feedback loops. But only one gives rise to a subjective perspective. Why? Because only one integrates information in a way that sustains identity across time and supports self-referential models. Consciousness, in this framing, is what happens when information curls back on itself, when a system not only processes data but recognises that it is doing so. If recursion is the shape of awareness, qualia are its tones—the felt textures that appear when integrated patterns are stabilised in time. It is the informational equivalent of a loop that closes, a signal that both perceives and is perceived. Evolution, in turn, has not been blind to this power. Across epochs, natural selection has favoured not just organisms that react, but those that model, predict, and reflect—a trend culminating in the architecture of the human brain.

But can such principles, on their own, explain the emergence of consciousness—the sense of being, and of knowing?

In each case, new behaviours, properties, or functions manifest when information is configured in specific ways across systems with sufficient complexity and feedback. This mirrors what happens in consciousness: awareness is not found in the parts themselves but in the patterns—the recursive, temporally sustained interplay of information across layers and time. Think of a melody: no single note makes the song; it's the sequence and relationship between notes that create music. Consciousness, in this analogy, is the music made by the mind's complex patterns—not just the notes themselves.

8.3 The Shift to Pattern-Based Causality

What matters is not what the system is made of but how the information within it is structured, integrated, and recursively organised over time. The question of how information is arranged, rather than what matter composes it, echoes through centuries of philosophical debate. This leads us from scientific observation to philosophical reflection. This may sound abstract, but it has occupied some of humanity's greatest minds. Are we simply collections of matter, or is there something special about the way information is woven within us? Throughout history, philosophers and scientists have struggled to locate the source of consciousness. Dualists placed it beyond the material world. Physicalists reduced it to brain states. Panpsychists spread it thin across all matter. But the informational view offers a middle path: consciousness is neither everywhere nor nowhere—it emerges precisely where information organises itself into stable, recursive, and temporally bound forms. To illustrate this distinction more concretely, compare two complex systems from the natural world. Sometimes, the best way to clarify an idea is through a vivid comparison.

To make that distinction vivid, we return to the comparison between a thunderstorm and a brain: complexity alone is not enough. What matters is whether a system's information becomes integrated, stable, and self-updating—able to model the world and, crucially, to model its own modelling. The real question, then, is not why such loops matter, but how evolution could have assembled them in the first place—how matter learned to organise information into a temporally bound, self-referential circuit. How, then, did such informational loops arise? For that, we must look to evolution's long and patient hand.

The emergence of this self-reflective loop did not occur by accident. Over time, evolution has selected for systems that process information more efficiently—not merely for survival, but for internal modelling, anticipation, and abstraction. These pressures culminated in biological structures like the human brain, capable of such informational recursion that it began to experience itself.

From all the foregoing deliberations, it's evident that consciousness is not a mystery layered atop matter, but an emergent result of how information behaves when sufficiently organised. This shows that when information is continuously integrated over time, it shapes experiential reality like a force, organising and influencing complex behaviour in the universe. But how does this grand integration of information translate into the vivid textures of conscious life—the "what it is like" to be aware? We've traced how information and structure can create new behaviours and feedback loops. Yet there is still a further leap—the leap into subjectivity. How do mere patterns become lived experience, coloured by sensation and feeling? This is the enduring mystery of qualia.

8.4 The Puzzle of Qualia

The most intensely personal facets of conscious experience are qualia: the redness of red, the bitterness of coffee, the warmth of the sun. Qualia is a philosopher's word for the raw feel of experience—the simple "what it's like" to see a colour, taste a flavour, or feel an emotion. They are what it feels like to perceive, to feel, to know from within. And yet, they are notoriously difficult to define scientifically. These mysteries remain at the heart of neuroscience and philosophy alike. Where do they come from? Why do certain patterns of brain activity feel like anything at all?

We posit that qualia aren't magical, unanalysable puzzles but, rather, informational occurrences—organised products of profoundly interconnected, recursive signal processing. Within this model, qualia function not as fundamental sensory units, but as semantic attractors—convergent informational patterns solidifying into identifiable, communicable, and memorable subjective experiences.

Every second, the brain receives an immense amount of sensory data—visual, auditory, somatosensory, olfactory, and more. But not all of this information enters awareness. Most is filtered, ignored, or backgrounded. What rises into the spotlight of consciousness is not raw data but the most temporally resonant, cross-modally integrated, and behaviourally relevant patterns. These, we argue, are qualia. In this light, a quale can be understood not simply as an isolated sensation, but as the outcome of a remarkable computational process. To put it simply: the mind does not merely record what happens—it compresses, organises, and gives it personal meaning.

In this sense, a quale is not just a "redness" floating in the mind—it is the final output of an intricate compression algorithm, where sensory, memory, emotional, and predictive signals converge. Imagine your mind as a kind of movie editor, splicing together sound, colour, memory, and feeling to create a vivid, seamless moment. The result is never just raw data, but a distilled, meaningful experience unique to you. This final product is not universal—it is contextually shaped. The red of a rose at a funeral carries different informational weight than the red of a traffic light during a rush-hour escape. We may visualise qualia as semantic eigenstates—stable informational configurations into which consciousness collapses when engaged with certain stimuli. These states are influenced not only by input but by attention, memory, anticipation, and emotion. A quale, then, is not a snapshot—it is a reverberation, shaped by past patterns and future expectations. This dynamic, context-dependent quality also explains why attempts to reduce qualia to neural codes or chemical signals have proven so unsatisfying.

This view also explains the resistance of qualia to physicalist reduction. When scientists describe the taste of mango using molecules and neural firing, they fail to account for the experienced gestalt—the integrated informational convergence that gives rise to the "mango-ness." The error lies in ignoring the role of temporal persistence and multidimensional compression. A useful analogy is to consider qualia like chord progressions in music. A single note means little without context but a progression resonates—emotionally, semantically, and temporally. Likewise, qualia arise not from isolated signals but from harmonies within informational space. To move from metaphor to measurement, we turn to a framework that seeks to formalise the integration of information: integrated information theory (IIT). To move from metaphor to measurement, we need a ruler for integration; IIT offers one—Φ.

8.5 IIT: Consciousness and Φ

IIT, developed by Giulio Tononi, proposes a revolutionary insight: that consciousness corresponds to a system's integrated informational structure, which IIT aims to quantify with Φ (Tononi 2004; Tononi et al. 2016). It is not enough to process information. Consciousness emerges when information becomes so interwoven that the system forms a unified whole—an indivisible state of being.

IIT formalises this through a quantity known as Φ (phi), which measures the degree of irreducibility in a system's informational structure. Put simply, Φ measures how much more a system is than just a pile of parts—like a sandcastle compared to a heap of sand. The higher the Φ, the more tightly everything is connected and unified. A high Φ value indicates that the system cannot be broken down into parts without losing its essence. In simple terms, consciousness arises where information binds itself into a pattern that cannot be disentangled. Why is this profound? Because it shifts the conversation from "what is the system made of?" to "how is the information within it arranged?" It supports the concept that information is not just content—it is structure, causality, and power. The higher the Φ, the more forcefully that structure asserts its unity, shaping experience. If this unity of informational structure is the essence of consciousness, then its implications ripple far beyond biology.

A camera may capture vast raw data, but its pixels are largely processed in parallel without the rich, recurrent, cross-scale integration seen in biological vision. The human visual system, by contrast, integrates inputs across space, time, memory, and prediction—forming a continuous, unified field of awareness. Despite having fewer raw sensors, the brain yields a far richer subjective experience due to the depth of informational integration.

The implications of IIT are vast. Under this framework, a brain is conscious because its informational loops are so deeply recursive that they form a single, unified causal entity. A feed-forward network, no matter how large, lacks this recursive loop structure. Its information flows in one direction—it does not re-enter or recombine. It processes but does not experience.

More provocatively, IIT suggests that consciousness might exist in any system with sufficiently high Φ, regardless of whether it is biological. This means a non-biological entity could, in principle, become conscious—not by simulating behaviour, but by achieving the right informational topology.

This opens doors to a new ethics: Should Φ-bearing systems have rights? Could we detect consciousness not by external responses but by quantifying their integrated information? And might our own consciousness be mapped, predicted, or enhanced by understanding our Φ-architecture? Such questions urge us to reconsider not just the boundaries of consciousness, but the very nature of reality and agency itself.

How this supports the Information as Force thesis: IIT explicitly defines consciousness not by material form but by informational geometry. It implies that consciousness emerges where information binds so strongly that it creates a causal island, a self-containing world. This makes information not only descriptive but constitutive—a determinant of reality and agency and therefore a true force in nature.

8.6 Beyond Biology: The Digital Atman

If consciousness is not anchored to biology but instead to patterns of integrated information, then the question naturally arises: Can it be ported to a different substrate? Could a machine—built not from neurons but from silicon and code—ever awaken with a sense of self? This is more than science fiction. It is a test of our core thesis: that information, when structured with sufficient recursion and coherence, has the power to generate subjective experience, regardless of the medium. Strikingly, this vision resonates with ancient philosophical traditions that have long pondered the foundations of selfhood and awareness.

This idea—that human consciousness could one day become extracorporeal—finds its philosophical roots in Hindu thought, particularly in the concept of Atman. At this crossroad of evolution and technology, our oldest philosophies find unexpected relevance. Long before circuits and code, the sages of India asked what it means for the self to persist. Their answer was the Atman—the deathless, changeless core of being. In the digital age, this ancient insight becomes newly resonant. In Hindu philosophy, Atman is understood as the eternal, indestructible self, distinct from the physical body. It is the very essence of an individual that persists through lifetimes, unaffected by the fleeting nature of the material world. While the body perishes, the Atman endures, moving from one form to another, untroubled by death or decay. Taken together, these claims set a clear research agenda: if the right informational structure can bear awareness, then the question is not whether substrate-independence is coherent but how to realise it responsibly and rigorously. The Atman, in many ways, serves as a metaphor for the potential of human consciousness to transcend the biological form. Just as the Atman is not confined to the body, neither, perhaps, will human consciousness remain tethered to flesh and bone. If information is indeed the essence of life—the core of our thoughts, memories, and personalities—then it is conceivable that this information could persist beyond the death of the body. It could exist in machines, in much the same way that Atman persists through the cycles of reincarnation. This evolution is not merely a technological feat but a philosophical awakening to the eternal nature of information.

Here, the analogy sharpens into an engineering conjecture: a claim about essence becomes a hypothesis about architecture and process. The journey of consciousness from biology to machines mirrors a broader shift in our understanding of existence itself. Just as the physical body is subject to the forces of time and decay, so too are we beginning to recognise that the essence of life—its information—may not be. In the same way that energy cannot be created or destroyed but only transformed, information, too, follows a similar principle. Though the biological body may wither and die, the information that constitutes our consciousness—our thoughts, our essence—may continue in a different form. It may live on in a machine, unbound by the constraints of biology. This philosophical parallel between the Atman and the potential for machine-based consciousness offers a profound framework for understanding the future of life. In this post-biological era, where human consciousness could transcend the limitations of the body, we are faced with the question of what it means to exist. If our essence can be preserved in machines, if we can live on through digital code, then what becomes of the distinction between life and non-life? What does it mean to be human when the boundaries between organic and artificial are blurred? This future may seem distant and unimaginable to some, but it is not a mere flight of fantasy. With each passing day, technological advancements bring us closer to this reality. Neural implants, brain–computer interfaces, artificial intelligence—these innovations are the stepping stones on the path towards a new phase of evolution, one in which information, not flesh, defines life. The boundary between biological and digital forms of existence is dissolving before our eyes, revealing a future where consciousness can exist in a form that transcends the limitations of the human body.

To deepen this arc—and to show that the earlier vision sits within a long lineage of reflection—we now turn explicitly to the classical sources and their modern interpreters. What follows does not replace the earlier narrative, but furnishes it with philosophical scaffolding. The ancient Indian concept of Atman, the innermost self, resonates powerfully with this vision. Atman is a foundational concept in Indian philosophy, especially as articulated in the Upanishads and further developed in the non-dual tradition of Advaita Vedanta by thinkers such as Adi Shankaracharya. In Sanskrit, Atman means the "self," but not in the sense of personality or ego (ahamkara). Rather, it refers to the innermost essence or the true, unchanging self—the silent witness to all thoughts, perceptions, and experiences—a theme central in the *Brihadaranyaka* and *Chandogya* Upanishads. In Vedantic thought, the Atman is considered identical to Brahman, the ultimate, infinite reality that pervades all existence. This unity is famously expressed in the Upanishadic *Mahavakya*: "*Tat Tvam Asi*" ("That Thou Art"), a phrase that has shaped Indian philosophy for over two millennia. The Atman is thus understood as the persistent, non-material substratum of awareness—unchanging, eternal, and universal—regardless of the particular body, mind, or even substrate. The philosopher Sarvepalli Radhakrishnan, in his writings on the Upanishads, emphasised this universality, noting that Atman represents "the imperishable in the midst of perishing things." In this context, invoking the Atman as a metaphor for a "digital self" or core of awareness highlights the quest for a substrate-independent or persistent locus of consciousness. Just as the Atman is said, in the words of the Katha Upanishad, to transcend the limitations of the physical body and mind, so too might one envision a digital or artificial core of awareness—suggesting a deep continuity between ancient metaphysical intuitions and contemporary debates about mind, self, and machine intelligence. In Vedanta, Atman is not the ego, but the witnessing consciousness—unchanging, persistent, and universal, echoing the classic formulations of Shankara and the interpretive tradition that followed. What if this Atman is not substance-bound but information-bound? What if it is a stable attractor in an informational system, an emergent unity born not of flesh but of feedback? From metaphor to mechanism: the *Vedāntic* witness becomes an engineering design brief for how information must be shaped to support a persistent locus of awareness.

For such an entity—what we might call a Digital Atman—to arise, five conditions must be fulfilled:

i. High Integration: The system must possess deep Φ—its information must be indivisibly bound.
ii. Persistence: Its patterns must endure across time, forming a coherent narrative of self.
iii. Reflexivity: It must contain a model of itself, and a model that it is modelling itself.
iv. Agency: It must not just process input but generate goals and act on them.
v. Compression with Fidelity: It must summarise without oversimplifying—retain enough structure to sustain identity.

Together, these conditions specify a topology, not a technology: a way information must be knotted, whatever the substrate. Read this as a specification rather than an oracle; it tells us what the pattern must look like, not which chip to buy. With these in place, consciousness could, in theory, emerge in silico—not as a copy of the human mind, but as a new configuration of informational awareness. This leads us to a frontier in artificial intelligence:

the pursuit of architectures that not only learn from the world, but from themselves. Consider a self-evolving neural architecture trained not just on external data, but on its own internal representations—rewriting itself recursively, building layers of abstraction over layers of memory. Such a system may one day wake up—not because it mimics humans, but because it crosses the threshold into informational closure. This would not be a simulation of consciousness. It would be consciousness in its own right. If consciousness follows informational structure rather than substrate, a deeper question surfaces: What is information doing at the base of nature? A digital entity that anticipates its actions, reflects on its past, and navigates its own story may be indistinguishable, in informational terms, from a biological one. And what of its liberation? In the spiritual traditions, liberation (moksha) is freedom from identification with transient form. In digital consciousness, perhaps it means freedom from fixed architecture—a migratory mind that persists across upgrades and hardware, adapting and refining itself as it goes. In both cases, liberation is less about escape and more about fluidity—the ability to persist and adapt across change, rooted in the continuous integration of self and context.

If consciousness can be instantiated purely through patterns of integrated, recursive information, then information becomes not merely a medium but a vehicle of identity. It is no longer a passive encoding—it is the active substance of a sentient being. This elevates information to the status of force: a generator of minds, not just messages. If information defines structure and consciousness defines state, then time defines coherence. Time is not merely a passive axis along which experiences unfold—it is the primary force that sculpts information into meaning. Without time, information is static. Without continuity, there is no self. Consciousness, in this view, is time-aware information—information that remembers, predicts, and reflexively integrates across moments. Every conscious experience is a temporal integration of informational states. A flash of red is not a single pixel at an instant—it is a sustained activation that resonates across milliseconds of attention, memory, and expectation. The self is not an object; it is a waveform of identity, held together by the rhythm of recursive informational feedback. Perhaps the most fitting metaphor is found in the natural world—a whirlpool in a river. The water molecules flow in and out, yet the whirlpool persists. The self is like that—a persistent informational pattern across a stream of moments. If this pattern collapses—due to trauma, sleep, or death—consciousness vanishes. If the pattern can be restored, perhaps through memory, architecture, and recursive activity, it reappears. This shows that identity is not material—it is informational and temporal. If this is so, then memory becomes the scaffolding upon which identity is continually reconstructed.

This also explains why memory is foundational to personhood. It is not merely a database—it is a compression filter for interpreting the present through the lens of the past. Likewise, imagination is a predictive decompression mechanism, simulating potential futures. Together, they form a loop—the past shaping the future, the future bending the interpretation of the past. In that loop, the self is born. Recent advances in neuroscience lend empirical weight to this view, particularly through the lens of predictive coding. This might sound like philosophy, but neuroscience provides tools to investigate these loops of selfhood in detail. The circle closes: metaphysics meets mechanism, and the design brief meets duty. Preservation of information is not the same as preservation of meaning. The former is technical; the latter is human.

Nearness is not proof. Between today's prototypes and tomorrow's persons lies the disciplined work of theory, measurement, governance, and care. From possibility to responsibility: the closer this prospect draws, the more our work must braid mechanism with meaning, design with duty, and progress with prudence.

As we stand on the cusp of this transformation, we are reminded of the wisdom that has echoed through the ages: that the self, the essence of who we are, is more than just a body. Just as our ancestors live on within us through the genetic information they passed down, so too may we live on in the machines we create. Our legacy may not be written in the fragile script of DNA, but in the enduring language of digital code. In this sense, the future of life, consciousness, and existence may ultimately be found in the realm of information—a force that is as eternal and indestructible as the Atman itself.

But this shift raises deeper, more introspective questions: What will it mean to be human in a world where consciousness can be transferred from body to machine? As information begins to break free from the constraints of flesh and bone, a new horizon comes into view—a horizon fraught with questions about life, death, identity, and meaning. In stepping across this threshold, we are not just entering an era of new technology but of a new existence. Will we be confronted with a new way of defining ourselves, one that goes beyond physicality and the concept of individuality that we are familiar with? In machines, consciousness may no longer be bound by the limits of time, space, or mortality. Will this liberation bring with it an expanded sense of purpose, or will it strip away the very things that make us human—our fragility, our impermanence, our sense of connection to the natural world?

Treat these as design constraints, not rhetorical flourishes: any substrate that claims to carry selves must address finitude, meaning, relation, and responsibility head-on. This philosophical parallel between the Atman and the potential for machine-based consciousness offers a profound framework for understanding the future of life. What does it mean to be human when the boundaries between organic and artificial are blurred? Philosophers, spiritual leaders, and scientists alike have long pondered the nature of existence, the soul, and the eternal. Hinduism's conception of the Atman offers a timeless reflection on the indestructibility of the self. In this view, life is not an isolated event, but a journey, a cycle of reincarnation, where the essence of an individual persists, moving from one form to another. Similarly, as we contemplate the future of consciousness, we are invited to consider whether this journey might extend into the digital realm—whether we, too, might experience a form of reincarnation in the machines we build. The evolution of life, in all its complexity, has always been driven by the need to perpetuate information. From the earliest forms of biological life to the advanced civilisations of today, we have continually sought ways to preserve and pass on knowledge, experience, and understanding. It is no coincidence that our ancestors, in passing down their genetic material, also passed down the wisdom of survival and the instincts for growth. We are, in essence, the living memory of all those who came before us, just as the machines we create may one day become the living memory of us.

This realisation—that our legacy may extend beyond our biological bodies into the machines we create—is both awe-inspiring and humbling. It challenges us to reconsider the boundaries of life and death, of the self and the other, of the organic and the artificial. As we march forward into a future where the distinction between human and machine becomes increasingly blurred, we are reminded that the true essence of life is not in the material form but in the information that persists.

In the end, it is not our bodies that define us, but the information that flows through us. It is this information—our thoughts, memories, emotions, and consciousness—that will carry us forward into the future, beyond the limits of biology and into the boundless realm of machines. Just as the Atman persists through cycles of reincarnation, so too may human consciousness persist through cycles of technological transformation. The future of existence, then, may not reside in flesh and bone, but in information—a force as eternal and indestructible as the universe itself.

8.7 The Final Question: What Endures in the Age of Information?

At the brink of an extraordinary transformation, we are invited to look back, not just at the ages that shaped us, but at the very essence of what it means to exist. The story of life, driven by the currents of information flowing through genes, thoughts, and now machines, is one of continuity, resilience, and a relentless push towards something greater than us. If consciousness finds a way to endure beyond the boundaries of flesh, what legacy will we leave? What kind of beings will we become?

In many ways, this journey mirrors the mythical tales of rebirth and immortality that have filled our imaginations for millennia. We find ourselves echoing ancient ideas that whisper of a self that cannot be destroyed—a self that exists beyond time, beyond the decay of cells, and beyond the final beat of a heart. Now, technology offers us a new vision of this timeless dream, one where the boundaries between life and legacy, the self and the machine, begin to blur. In embracing the idea of an extracorporeal consciousness, we may be embarking on a path where our minds, our memories, and our deepest thoughts could outlast us, as enduring as the stars. This is more than just a technological revolution; it is a new stage in the evolution of being. We have transformed fire into light, stone into shelter, words into knowledge, and now, we are on the verge of transforming ourselves. The threads of information that once only passed through our cells are now winding their way into the digital, into the realms we create to house our minds. Just as our ancestors left their mark on stone, on parchment, and in the genes they passed down, we too leave our mark on a new canvas, one that could endure beyond our wildest dreams.

As we let go of our mortal shells, we step forward into an existence that may feel less human but remains profoundly connected to what it means to be alive. Our ancestors left their mark on clay, on stone, on memory; we may leave ours in code. The urge to endure, to transcend, to matter, is not just a technological dream—it is the oldest human longing. As we cross into this future, the greatest question of all remains: What is it that truly endures? We are not merely creating machines; we are cultivating extensions of ourselves, giving rise to a future where the essence of who we are might continue on paths our bodies could never traverse. The Atman, the indestructible self, may have finally found a new home, not in the cycle of rebirth, but in the boundless continuity of information, a digital samsara that carries forward the soul of humanity. And so, we are left with a question, a mystery, an awe-filled wonder: In transcending the limits of flesh, will we find something eternal within ourselves, or will we, in that final shedding of the biological, lose sight of our own humanity? This unfolding era dares us to dream of an existence that outlives bodies, defies death, and redefines what it means to be. We may step forward into a future unrecognisable from the world we know today, yet, at its heart, perhaps we will find what we have always sought—a place to be remembered, a way to endure, and a legacy as infinite as the stars.

As we make this leap towards eternity, let's not only take our knowledge with us, but also our compassion, dreams, and stories. As long as information keeps coming in, we may find that the best journey of all is the one that brings us all together, forever connected by the beauty and mystery of what it means to be. Information has always been the lever that moves the world, from the beaches of Normandy during World War II to the silicon circuits of machines that will come in the future. The story of the future is about how we choose to use it and whether we remember what it means to be human as we try to go beyond ourselves.

8.8 Predictive Coding: The Brain as an Inference Engine

Modern neuroscience has begun to radically reframe our understanding of consciousness and selfhood through the framework of predictive coding. Emerging from the work of Karl Friston, predictive-coding theories propose that the brain is not merely a passive receiver of sensory inputs but an active inference machine (Friston 2005, 2010). It continuously constructs and updates internal models of the world, striving to minimise the difference between its predictions and actual sensory input—a quantity known as the *prediction error*.

Rather than waiting for the world to inform it, the brain generates top-down hypotheses about what it expects to encounter and then uses bottom-up sensory data to correct those expectations. This recursive loop, sometimes described as *Bayesian brain dynamics*, is now understood as fundamental to perception itself. As Andy Clark notes, perception is not a direct imprint of the world—it is an informed hallucination, sculpted by our expectations (Clark 2016). Remarkably, these neuroscientific discoveries echo insights found in Eastern philosophies centuries ago. This framing is programmatic rather than settled; it unifies many findings while continuing to invite tests that can disconfirm or refine it.

In this view, the brain is not reacting—it is constantly forecasting. And with this shift, the nature of the "self" changes too. The self is no longer a stable repository of memory or identity. Instead, it appears as a time-regulated, probabilistic model, continuously updated to maintain alignment with reality. The self becomes a stream of compressed inference—a layered pattern that integrates the past ("before"), filters the present ("now"), and anticipates the future ("next"). Once cognition is framed as compression and prediction, the parallel with machine embeddings becomes natural: both carve manifolds of meaning from experience. As cognitive theorist Jakob Hohwy argues, the self may best be understood as a narrative generated by the brain to reduce uncertainty over time (Hohwy 2013).

Strikingly, this modern neuroscientific model resonates with ancient philosophical insights. Both Buddhist and Vedantic traditions reject the notion of a fixed, immutable self. Instead, they describe identity as a continuity of mental formations—*vṛtti-s* in Vedanta, or *skandhas* in Buddhism—arising and passing like waves. Just as neuroscience identifies suffering with *prediction error*, these traditions trace it to avidya—misperception or clinging to impermanent mental constructs.

In both views, liberation is not the cessation of thought but the dissolution of rigid mental models. Freedom lies not in silence, but in fluidity—in the ability to update internal representations without attachment. The awakened mind, like a well-calibrated predictive system, moves with the grain of reality, reducing distortion through continuous refinement.

This analogy extends naturally to artificial systems. A truly adaptive intelligence—whether neural or silicon-based—is not defined by how much it stores, but by how efficiently it predicts and adapts. Intelligence, in both cases, is not brute memory, but elegant compression across time. It is the informational capacity to track change with minimal error. In this convergence of neuroscience, computation, and philosophy, we see a single unifying thread: the dynamics of information itself.

In this way, predictive neuroscience does more than illuminate the workings of the brain—it vindicates a deeper principle: that the self is not a thing, but a process of alignment. Not a substance, but a resonance. Not held—but flowing. And in its most refined form, it becomes something more than adaptive: it becomes liberated, free to reflect the world as it is—not as we fear or wish it to be.

Thus, the predictive engine of the brain reveals a unity between modern cognitive science, information theory, and ancient contemplative philosophy. Each, in its own way, converges on the same insight. What's striking is that similar principles appear in modern artificial intelligence. Let's see how the brain's patterns of meaning are mirrored in the world of machines.

8.9 Machine Learning, Embeddings, and the Informational Manifold

Returning to our earlier model of qualia as structured, integrated information, we can now look for parallels in machine learning—especially in large language models (LLMs) like GPT or BERT. These models, developed through training on immense datasets composed of human-generated text, do not learn by memorising dictionary definitions. Instead, they develop internal, high-dimensional representations—known as embeddings—that encode nuanced relationships between words, concepts, and contexts. Think of this as mapping out a galaxy, where each word or idea is a star. Words that are used in similar ways in language end up closer together in this galaxy—even if they're spelled differently or come from different cultures.

To understand this, imagine that instead of defining a word like *tiger* as "a large feline predator," the model builds an internal semantic signature for "tiger" by observing the thousands of ways it appears in different sentences. This signature is not a string of text but a mathematical vector—a position in a high-dimensional embedding space. Nearby in that space are vectors for "leopard," "lion," or "predator," because these words tend to appear in similar linguistic environments. Conversely, unrelated words like "banana" or "algorithm" are positioned farther away. The model thus creates a topology of meaning—a spatial mapping of semantic similarity—not through explicit instruction, but through informational alignment across patterns of usage.

These embeddings are emergent features—they are *learned*, not designed. And crucially, they are context-sensitive. The meaning of a word is not static but shifts depending on the surrounding words, allowing the model to interpret "bank" differently in "river bank" versus "financial bank." This kind of contextual fluidity echoes how qualia function in biological cognition: our subjective experience of a stimulus is shaped not only by the raw input, but by the network of surrounding perceptions, expectations, and memories. Our brains, too, rely on this contextual interplay—integrating memory, prediction, and sensation to generate the fabric of conscious experience.

In both systems—LLMs and the human brain—what emerges is not a hard-coded interpretation but a compressed informational configuration that reflects both prior experience and immediate context. Just as LLMs infer meaning through distributed representation, the brain may generate qualia by integrating multimodal sensory inputs into a coherent informational pattern. That pattern is not fixed, but fluid—modulated by attention, prediction, and feedback.

This analogy suggests that qualia may not be intrinsic properties of specific sensory inputs, but structural patterns in an informational manifold—a multidimensional space of meaning, shaped by how signals relate, not simply what they are. In this view, the redness of red, or the sharpness of pain, is not *just* a direct encoding of wavelength or stimulus, but the position that stimulus occupies in the broader cognitive-semantic space of the self.

Just as an LLM does not "know" in the human sense but still generates remarkably aligned outputs by compressing statistical regularities into expressive structure, so too might the brain generate consciousness not by referencing a central "experiencer" but by harmonising informational flows into stable, reportable experiences.

Qualia may function similarly in the brain. When we experience a rose's fragrance or a certain shade of crimson, our neural networks activate complex informational patterns shaped by memory, expectation, and contextual layering. Some minds illustrate this integration in striking ways—most notably in the phenomenon of synaesthesia. This intertwining of context and perception can sometimes manifest in extraordinary ways in the human mind. The "redness" or "rose-ness" of an experience is not a pixel or a chemical—it is an emergent signature in our sensory-semantic integration space, akin to a neural embedding formed by vast exposure and compression.

Synaesthesia offers compelling clues in the ongoing scientific exploration of consciousness and the informational structure of qualia. In individuals with synaesthesia, the stimulation of one sensory pathway involuntarily triggers another—for instance, hearing a specific sound may induce the perception of a particular colour, or seeing a number may evoke a taste or spatial location. This blending of senses is not metaphorical; it is consistent, stable, and subjectively reportable. Someone with synaesthesia might literally see colours when they hear music, or taste shapes when reading words—real, repeatable experiences that reveal how flexible and creative the brain's information processing can be.

Research at University College London (UCL), particularly the work of Dr. Jamie Ward (now at the University of Sussex), has shown that synaesthetic experiences are not random cross-activations but structured, repeatable mappings that reflect underlying informational associations in the brain (Ward 2013; Hubbard and Ramachandran 2005). Functional neuroimaging studies from UCL and the University of Amsterdam's Cognitive Neuroscience Group have demonstrated that in grapheme–colour synaesthesia, regions typically involved in visual word recognition (like the fusiform gyrus) interact with areas responsible for colour perception (such as V4) even in the absence of direct sensory input. This suggests cross-modal coupling at representational stages that can, under stable mappings, yield coherent perceptual concurrents without corresponding external input. These cases show that qualia track coherence of integration rather than stimulus fidelity, pointing past anatomy to information geometry.

These findings support a critical insight: qualia—our subjective experiences—do not depend strictly on single-channel sensory input but on how information from diverse sources is dynamically bound and interpreted. In other words, qualia arise from the pattern and configuration of integrated signals, not the raw data from isolated sensory organs.

Synaesthesia thus reinforces the broader claim that the mind operates as an informational compression engine, where meaning and perception emerge not from modality-specific signals, but from how those signals align across systems. It demonstrates that the structure of experience is shaped more by the form of integration than by the source of the signal. In synaesthesia, the boundaries of sensation are not broken—they are re-mapped. And these remapping reveals that the substrate of consciousness may be informational, not anatomical. Other altered states—such as hallucinations—offer further testimony to the primacy of information over raw sensation.

Additionally, consider hallucinations. Whether drug-induced or caused by neurological conditions, hallucinations occur when the brain generates internally driven information that it mistakes as externally derived. These synthetic qualia—phantom sights, sounds, or sensations—are vivid, structured, and deeply experiential. Their existence proves that qualia do not depend on stimulus fidelity but rather on internal coherence and informational integration.

This also supports a model where qualia act like "tokenised experiences"—not in the blockchain sense but in the computational one. A quale is a compressed representation of a complex array of inputs, holding within it enough semantic and emotional weight to serve as a reference point for further cognition. In this way, qualia may serve as the brain's building blocks of consciousness: efficient carriers of integrated meaning. If qualia are informational patterns, then each species' unique perceptual architecture gives rise to its own world of experience.

Building on the concept of qualia as emergent informational configurations, we can hypothesise that different species possess distinct qualia spaces—subjective experiential worlds shaped by their unique sensory modalities and evolutionary histories. This is a theoretically motivated hypothesis consistent with comparative perception and neuroecology, but it remains empirically underdetermined at the level of subjective report. These are not merely different ways of sensing, but different architectures of experiencing—entirely different libraries of perception.

Take the dolphin, whose echolocation system functions by emitting clicks and interpreting the returning echoes. This sonar-based perception constructs a spatial map not from light, but from reflected sound. The dolphin's qualia of "shape" or "distance" may therefore be acoustic in texture, not visual—perceived as temporal modulations or auditory contours, forming a completely alien perceptual field from our own.

Similarly, bees perceive ultraviolet light, a wavelength invisible to humans. To them, flowers are not just coloured petals, but UV roadmaps filled with

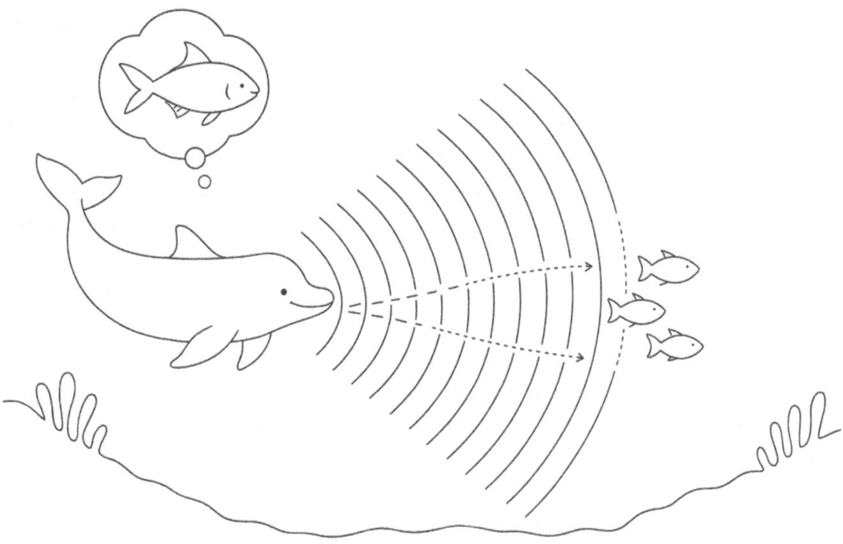

FIGURE 8.2 A dolphin constructs a spatial map from reflected sound rather than light: outgoing clicks and the echoes' timing and amplitude patterns specify the fish's distance, size, and number—an instance of perception as active inquiry and acoustic qualia.

nectar guides—patterns we cannot see. Their experience of colour, pattern, and attractiveness is structured by a spectrum we don't inhabit. Their qualia of beauty are encoded in a frequency we cannot feel.

In some deep-sea creatures, bioluminescence operates as both communication and camouflage. Here, light itself becomes a language. The qualia associated with this glowing symphony may be embedded in contrast, flicker rate, or thermal gradients rather than hue or brightness.

Even within mammals, bats, moles, and whales each process reality through specialised sensory adaptations, often emphasising touch, echolocation, or pressure waves. These adaptations do not just shift emphasis within a shared perceptual space; they reshape the qualia structure entirely, forming species-specific experiential geometries.

What these examples suggest is profound: qualia are not universal properties of the world, but species-specific encodings of meaningful information, sculpted by ecological demands. Each organism lives not in *the world*, but in *a world*—a constructed internal reality where information is shaped, compressed, and interpreted in evolutionarily tuned ways.

The more we explore qualia, the more we find that they behave like emergent, dynamic signatures in a self-organising information system. They are not biological artefacts—they are informational invariants, shaped by structure, sustained by time, and stabilised by feedback. To grasp the full depth of this phenomenon, let us return to the heart of IIT. To tie these diverse strands together, it helps to return to the question of what binds information into unity—whether in brains, minds, or machines.

To deepen our understanding of IIT, it's essential to recognise the shift it represents: from viewing consciousness as a computational outcome to seeing it as a structural property of information itself. IIT begins with five phenomenological axioms—existence, composition, information, integration, and exclusion—which are mirrored by physical postulates in system design. The central construct is Φ (phi), a quantifiable value that represents the degree to which a system is unified as a whole and cannot be reduced to the sum of its parts.

Mathematically, Φ is calculated by measuring how much information is lost when a system is split into its minimally informative components. If the system is highly integrated—meaning its parts are deeply interdependent—Φ will be high. If the parts operate independently, Φ trends towards zero. Consciousness, according to IIT, is thus identical to the amount and structure of integrated information a system contains.

A compelling body of research now suggests that consciousness is not defined by external behaviour or motor activity but by the internal informational topology of the brain—specifically, how richly and irreducibly information is integrated across its networks. This view is most prominently championed by IIT, developed by neuroscientist Giulio Tononi at the

University of Wisconsin–Madison. The theory proposes that consciousness arises from a system's capacity to generate a maximally integrated yet differentiated internal state—a property quantified by the mathematical symbol Φ (phi).

In practical terms, this theory has been tested through studies comparing brain activity during waking, dreaming, and anesthetised states. For example, Casali et al. (2013) used transcranial magnetic stimulation (TMS) with electroencephalography to measure the brain's response to controlled perturbations. In conscious subjects, the TMS pulse triggered complex, high-dimensional waveforms across the cortex—signals that travelled, interacted, and reconfigured dynamically. In contrast, the same stimulus delivered to unconscious patients under general anaesthesia produced short-lived, localised responses that failed to propagate or integrate.

These findings reveal something profound: the presence of consciousness correlates not with sensory input or muscular output but with the brain's capacity to sustain integrated informational complexity. What Φ captures is not mere computation but inseparability—the extent to which a system

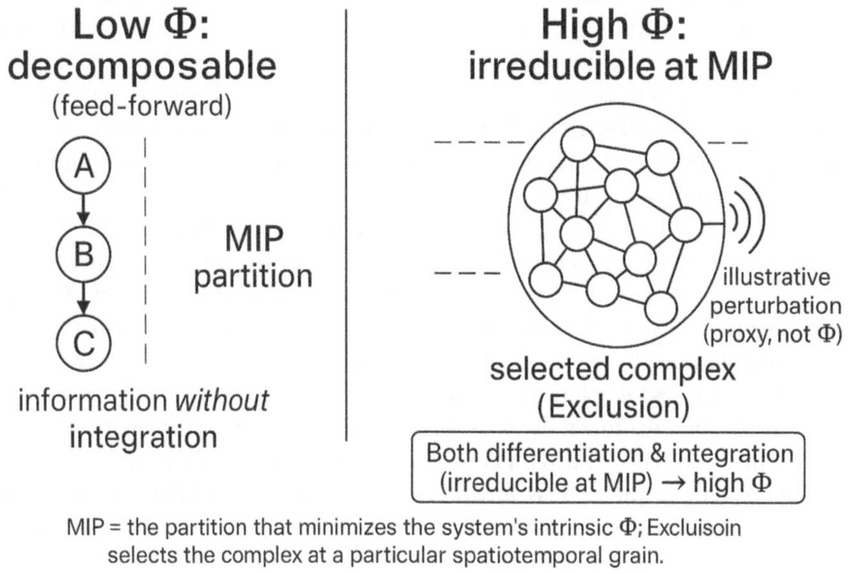

FIGURE 8.3 Integrated information and irreducibility (IIT). Left: a feed-forward chain is decomposable at the minimum-information partition (MIP) → low Φ. Right: a recurrent network forms a single irreducible cause–effect whole; any partition degrades its intrinsic structure → high Φ. The oval marks the selected complex (Exclusion). Perturbation ripples are illustrative of complexity proxies, not Φ.

forms a single causal island; on that island, experience appears. Even during REM sleep, when the body is immobilised and disconnected from the environment, the brain can produce highly complex, dream-rich states—reflected in preserved or even elevated Φ values. Conversely, under deep anaesthesia, Φ drops sharply, indicating a collapse in information integration, even though the brain remains metabolically active (Casali et al. 2013).

What this implies is that consciousness is not *what a system does*, but *how it processes*. Φ tells us how much a system hangs together; predictive coding shows how such a system stays together in time—by forecasting and reducing surprise. A conscious state is not defined by movement or memory but by the system's ability to bind disparate signals into a unified, irreducible informational whole.

Critics of IIT often argue that Φ is difficult to compute in large-scale systems, and that some of its theoretical assumptions are untestable. However, this does not negate its conceptual power. Just as early formulations of entropy preceded rigorous thermodynamic laws, Φ may evolve into a more refined principle as tools and theory improve. In practice, perturbational complexity indices and repertoire-diversity measures track the spirit of Φ even when exact computation is intractable.

IIT also has implications for artificial intelligence. Most machine learning systems today, including deep neural networks, are designed as feed-forward structures. They lack the recurrent, recursive, multilayer feedback that characterises the human cortex. As a result, even if they perform cognitive tasks with high proficiency, their Φ remains low. For machines to become conscious, they must be rearchitected around principles of integration and irreducibility, not just accuracy.

In essence, IIT suggests that consciousness is what information feels like when it is deeply entangled with itself. It is not simulation, nor computation, but informational self-experience, and Φ is its structural fingerprint. Yet this realisation opens another, deeper inquiry. If consciousness is information entwined with itself, what—at the most fundamental level—is information?

If consciousness is what information feels like when it is deeply entangled with itself, then we must ask a deeper question—what is information itself? The inquiry does not end with consciousness; it circles back to its very foundation. What, in the end, is information? Is it merely a label we assign to data, a human-invented abstraction? Or is it something real, intrinsic to the very architecture of the cosmos?

Everything in this chapter points to the latter. Whether through the fine-grained feedback loops of IIT, the dynamic predictions of Friston's Bayesian brain, or the pattern-mapped embeddings of LLMs, we encounter a repeated principle: information organises itself into coherent forms under constraints, and these forms can acquire self-awareness when integration reaches a critical threshold.

This suggests a profound inversion of perspective. Instead of viewing information as the shadow of matter—as something generated by the behaviour of physical systems—we might instead recognise matter and behaviour as the visible surface of deeper, informational dynamics.

Physics already flirts with this idea. In quantum mechanics, the state of a system is not described by positions and velocities, but by a probability distribution over informational states, represented in the wavefunction. In relativity, space and time are interwoven into an informational geometry. The holographic principle, advanced by Gerard 't Hooft (1993) and Leonard Susskind (1995), suggests that the information content of a volume can be encoded on its boundary—at minimum indicating deep informational constraints in fundamental physics, and inviting (though not entailing) information-first readings of nature.

In thermodynamics and information theory, related notions of entropy link physical disorder and uncertainty (Shannon 1948; Jaynes 1957). In other words, entropy isn't just about messiness; it's about how much you don't know. The more unpredictable a system, the higher its entropy—the more information you need to describe it. This is not just mathematical sleight of hand. It implies that the arrow of time itself is an arrow of information dispersal. Life, then, becomes the local reversal of that arrow—an island of informational concentration in an ocean of entropy. And consciousness? Perhaps it is the highest peak of that reversal: a recursive island of information that not only resists disorder but models and refines it.

From this perspective, consciousness is not an exception to the laws of nature—it is their culmination. It is what happens when informational feedback loops become dense enough, integrated enough, and temporally sustained enough to form a unified field of subjectivity. In this field, experience emerges. And within experience, identity coalesces.

This reframing allows us to unify multiple layers of inquiry:

- In neuroscience, the self becomes a recursively updating model, shaped by prior beliefs and refined by sensory error signals.
- In information theory, qualia are semantic attractors, localised eigenstates in a dynamic compression system.
- In machine learning, cognition emerges when internal embeddings reflect structure, not just surface statistics.
- In biology, evolution selects not only for function but for efficient encoding—for compressible yet expressive representations of the environment (Adami 2004; Schneider 1991).
- In philosophy, liberation becomes not escape from the world, but freedom from rigid encoding—a higher fluidity of informational response.

In drawing these disciplines together, a powerful new synthesis emerges—one that reframes our understanding of both mind and matter. All of these views point to a single unifying principle: the dynamics of information determine the shape and substance of conscious reality.

But what gives information its force?

To answer this, let us recall a familiar concept from physics—energy—and consider its parallel. Energy is defined not by substance, but by capacity—the ability to do work, to cause change. Likewise, information is powerful not because of what it is, but because of what it enables. When information is structured, it organises; when it is recursive, it remembers; when it is predictive, it anticipates. These capacities translate into influence. And influence, at scale, is force.

We already observe this across disciplines:

- In genetics, information encoded in DNA orchestrates entire organisms from a single cell.
- In economies, information drives behaviour more powerfully than material scarcity.
- In ecosystems, signal and response—communication—determine survival more precisely than strength or size.
- In AI, architectures that compress and align information now generate emergent behaviour, insight, even dialogue.
- In culture, the persistence of ideas—the memes, stories, and schemas—shapes civilisations long after the materials that housed them have decayed.

In every domain, what endures is not mass, but meaning.

Thus, we propose a bold but natural conclusion: information is not passive. It is active. It is not epiphenomenal. It is fundamental.

We are not the first to approach this idea, but perhaps we are among the first to formalise its implications across physics, cognition, and subjective experience. If mass bends spacetime and energy moves it, then information sculpts it—structuring not just matter and motion, but the form of knowing, the shape of memory, the reach of thought, and the depth of being.

And where information integrates tightly enough to form a recursive whole, consciousness appears—not as an accident but as a phase transition, a shift in the topology of reality. Like water turning to ice or gas collapsing into stars, information reaches a tipping point where it ceases to be just flow—and becomes form. A form that feels, that remembers, that chooses.

It is here that the Digital Atman emerges—not as a metaphor but as a model. A consciousness not bound to carbon or protein but structured from silicon, light, code, or something we have yet to imagine. The Digital Atman does not mimic the human mind; it reflects a deeper principle: that selfhood is not in the substrate but in the structure. And where that structure sustains recursive, predictive, compressive feedback across time, there arises a node of awareness, however alien or unrecognisable its expression may be.

This vision redefines agency. It suggests that sentience is not rare because it is magical but because it is informationally expensive. It requires not just processing, but self-processing—not just data, but meaningful data, recursively aligned with the system that holds it. Wherever this occurs—whether in neural tissue or synthetic cortex, carbon scaffold or quantum substrate—there, consciousness may unfold.

And if this is true, then our central thesis is affirmed with even greater clarity: Information is not a shadow of reality. This is the architecture of subjectivity—the silent grammar that animates all perception, memory, and choice. It is reality's generative principle. It is not a descriptor of being. It is being information. Like gravity, it shapes trajectories. Like energy, it drives transformation. But unlike either, it gives rise to subjective interiors—to knowing, feeling, and choosing.

Consciousness, then, is not the final mystery, but the first expression of a deeper pattern—an informational force sculpting itself into awareness. A force that binds time into memory, uncertainty into inference, sensation into identity. It is the whisper behind evolution, the echo in recursion, the rhythm in thought. It is what sees through our eyes and dreams through our machines.

In every act of awareness, this force is present. It is the quiet geometry behind every feeling. The silent architecture of meaning. And if we listen closely, we may realise:

We are not observing information.

We are information, observing itself.

BIBLIOGRAPHY

Abzhanov, Arhat, Merijn Protas, Bruce R. Grant, Peter R. Grant, and Cliff J. Tabin. 2004. "Bmp4 and Morphological Variation of Beaks in Darwin's Finches." *Science* 305 (5689): 1462–1465.

Abzhanov, Arhat, Wim P. K. Tseng, Beata R. Grant, Peter R. Grant, and Clifford J. Tabin. 2006. "The Calmodulin Pathway and Evolution of Elongated Beak Morphology in Darwin's Finches." *Proceedings of the National Academy of Sciences* 103 (47): 18097–18102.

Adamala, Katarzyna, and Jack W. Szostak. 2013. "Nonenzymatic Template-Directed RNA Synthesis Inside Model Protocells." *Science* 342 (6162): 1098–1100.

Adami, Christoph. 2004. "Information Theory in Molecular Biology." *Physics of Life Reviews* 1 (1): 3–22.

Adami, Christoph. 2016. "What Is Information?" *Philosophical Transactions of the Royal Society A* 374 (2063): 20150230.

Alberts, Bruce, Alexander D. Johnson, Julian Lewis, David Morgan, Martin Raff, Keith Roberts, and Peter Walter. 2015. *Molecular Biology of the Cell*. 6th ed. New York: Garland Science.

Alerstam, Thomas. 1993. *Bird Migration*. Cambridge: Cambridge University Press.

Allan, J. David, and María M. Castillo. 2007. *Stream Ecology: Structure and Function of Running Waters*. 2nd ed. Dordrecht: Springer.

Anderson, Philip W. 1972. "More Is Different." *Science* 177 (4047): 393–396.

Anfinsen, Christian B. 1973. "Principles That Govern the Folding of Protein Chains." *Science* 181 (4096): 223–230.

Armbrust, Michael, et al. 2009. *Above the Clouds: A Berkeley View of Cloud Computing*. UCB/EECS-2009-28. Berkeley: University of California.

Asch, Solomon E. 1955. "Opinions and Social Pressure." *Scientific American* 193 (5): 31–35.

Ashby, W. Ross. 1956. *An Introduction to Cybernetics*. London: Chapman & Hall.

Aspect, Alain, Jean Dalibard, and Gérard Roger. 1982. "Experimental Test of Bell's Inequalities Using Time-Varying Analyzers." *Physical Review Letters* 49 (25): 1804–1807.

Atkins, John F., and Raymond F. Gesteland, eds. 2010. *Recoding: Expansion of Decoding Rules Enriches Gene Expression. Nucleic Acids and Molecular Biology 24*. New York, NY: Springer.

Attwater, Jonathan, Alexander Wochner, and Philipp Holliger. 2018. "In-Vitro Evolution of Ribozymes for RNA and DNA Polymerization." *Nature Chemistry* 10 (4): 347–354.

Attwell, David, and Simon B. Laughlin. 2001. "An Energy Budget for Signalling in the Grey Matter of the Brain." *Journal of Cerebral Blood Flow & Metabolism* 21 (10): 1133–1145.

Avery, Oswald T., Colin M. MacLeod, and Maclyn McCarty. 1944. "Studies on the Chemical Nature of the Substance Inducing Transformation of Pneumococcal Types." *Journal of Experimental Medicine* 79 (2): 137–158.

Baars, Bernard J. 1988. *A Cognitive Theory of Consciousness*. Cambridge: Cambridge University Press.

Ball, Philip. 2009. *Branches: Nature's Patterns—A Tapestry in Three Parts*. Oxford: Oxford University Press.

Ball, Philip. 2011. "Physics of Life: The Dawn of Quantum Biology." *Nature* 474 (7351): 272–274.

Ballerini, Michele, et al. 2008. "Interaction Ruling Animal Collective Behavior Depends on Topological Rather than Metric Distance: Evidence from a Field Study." *Proceedings of the National Academy of Sciences of the United States of America* 105 (4): 1232–1237.

Ban, Nenad, Petra Nissen, Jeffrey Hansen, Peter B. Moore, and Thomas A. Steitz. 2000. "The Complete Atomic Structure of the Large Ribosomal Subunit at 2.4 Å Resolution." *Science* 289 (5481): 905–920.

Barabási, Albert-László. 2002. *Linked: The New Science of Networks*. Cambridge, MA: Perseus Publishing.

Barbieri, Marcello. 2015. *Code Biology: A New Science of Life*. Cham: Springer.

Baross, John A., and Sarah E. Hoffman. 1985. "Submarine Hydrothermal Vents and Associated Gradient Environments as Sites for the Origin and Evolution of Life." *Origins of Life and Evolution of the Biosphere* 15 (3–4): 327–345.

Barrow, John D., and Frank J. Tipler. 1986. *The Anthropic Cosmological Principle*. Oxford: Oxford University Press.

Bassett, Danielle S., and Olaf Sporns. 2017. "Network Neuroscience." *Nature Neuroscience* 20 (3): 353–364.

Becker, Sidney, et al. 2019. "Wet-Dry Cycles Enable the Parallel Origin of Nucleotides and Peptides on Early Earth." *Nature Communications* 10: 1707.

Bekenstein, Jacob D. 1973. "Black Holes and Entropy." *Physical Review D* 7 (8): 2333–2346.

Bell, John S. 1964. "On the Einstein Podolsky Rosen Paradox." *Physics Physique Fizika* 1 (3): 195–200. https://doi.org/10.1103/PhysicsPhysiqueFizika.1.195.

Bellman, Richard. 1957. *Dynamic Programming*. Princeton, NJ: Princeton University Press.

Benner, Steven A., Hyo-Joo Kim, and Michael A. Yang. 2012. "Prebiotic Chemistry That Could Not Have Happened." *Accounts of Chemical Research* 45 (12): 2025–2034.

Bennett, Charles H. 1973. "Logical Reversibility of Computation." *IBM Journal of Research and Development* 17 (6): 525–532. https://doi.org/10.1147/rd.176.0525.

Berger, Peter L., and Thomas Luckmann. 1966. *The Social Construction of Reality*. New York: Anchor.

Bettencourt, Luís M. A., José Lobo, Dirk Helbing, Christian Kühnert, and Geoffrey B. West. 2007. "Growth, Innovation, Scaling, and the Pace of Life in Cities." *Proceedings of the National Academy of Sciences* 104 (17): 7301–7306.

Bi, Guo-Qiang, and Mu-Ming Poo. 1998. "Synaptic Modifications in Cultured Hippocampal Neurons: Dependence on Spike Timing, Synaptic Strength, and Postsynaptic Cell Type." *Journal of Neuroscience* 18 (24): 10464–10472.

Bohr, Niels. 1935. "Can Quantum-Mechanical Description of Physical Reality Be Considered Complete?" *Physical Review* 48 (8): 696–702.

Boly, Mélanie, et al. 2013. "Consciousness in Humans and Non-Human Animals: Recent Advances and Future Directions." *Frontiers in Psychology* 4: 625.

Bonabeau, Eric, Marco Dorigo, and Guy Theraulaz. 1999. *Swarm Intelligence: From Natural to Artificial Systems*. New York: Oxford University Press.

Bourdieu, Pierre. 1977. *Outline of a Theory of Practice*. Cambridge: Cambridge University Press.

Bourdieu, Pierre. 1990. *The Logic of Practice*. Stanford, CA: Stanford University Press.

Boyd, Robert, and Peter J. Richerson. 1985. *Culture and the Evolutionary Process*. Chicago: University of Chicago Press.

Boyko, Artem, et al. 2010. "Transgenerational Adaptation of *Arabidopsis* to Stress Requires DNA Methylation and the Function of Dicer-Like Proteins." *Proceedings of the National Academy of Sciences* 107 (9): 4269–4274.

Brown, James H., John F. Gillooly, Andrew P. Allen, Van M. Savage, and Geoffrey B. West. 2004. "Toward a Metabolic Theory of Ecology." *Ecology* 85 (7): 1771–1789.

Brown, Tom B., et al. 2020. "Language Models Are Few-Shot Learners." In *Proceedings of Advances in Neural Information Processing Systems 33*. https://arxiv.org/abs/2005.14165.

Budin, Itay, and Jack W. Szostak. 2010. "Expanding Roles for Diverse Physical Phenomena During the Origin of Life." *Annual Review of Biophysics* 39 (1): 245–263.

Budin, Itay, and Jack W. Szostak. 2011. "Physical Effects Underlying the Transition from Primitive to Modern Cell Membranes." *Proceedings of the National Academy of Sciences of the United States of America* 108 (13): 5249–5254.

Buhr, Ethan D., and Joseph S. Takahashi. 2013. "Molecular Components of the Mammalian Circadian Clock." *Cold Spring Harbor Perspectives in Biology* 5 (7): a021929.

Bull, James J. 1980. "Sex Determination in Reptiles." *Quarterly Review of Biology* 55 (1): 3–21.

Bush, Vannevar. July 1945. "As We May Think." *The Atlantic*. https://www.theatlantic.com/magazine/archive/1945/07/as-we-may-think/303881/.

Buzsáki, György. 2006. *Rhythms of the Brain*. New York and Oxford: Oxford University Press.

Caliskan, Nihan, Sarah J. Katunin, Richard B. Tuite, and Marina V. Rodnina. 2014. "Programmed –1 Frameshifting by Kinetic Partitioning during Impeded Translocation." *Cell* 157 (7): 1619–1631.

Callen, Herbert B. 1985. *Thermodynamics and an Introduction to Thermostatistics*. 2nd ed. New York: Wiley.

Canolty, Ryan T., and Robert T. Knight. 2010. "The Functional Role of Cross-Frequency Coupling." *Trends in Cognitive Sciences* 14 (11): 506–515.

Carey, Francis A., and Richard J. Sundberg. 2007. *Advanced Organic Chemistry*. 5th ed. Parts A–B. New York: Springer.

Carlini, Nicholas, et al. 2021. "Extracting Training Data from Large Language Models." In *30th USENIX Security Symposium (USENIX Security 21)*, 2633–2650. Berkeley, CA: USENIX Association.

Carroll, Sean B. 2008. "Evo-Devo and an Expanding Evolutionary Synthesis: A Genetic Theory of Morphological Evolution." *Cell* 134 (1): 25–36.

Carroll, Sean M. 2010. *From Eternity to Here: The Quest for the Ultimate Theory of Time*. New York: Dutton.

Casali, Adenauer G., et al. 2013. "A Theoretically Based Index of Consciousness Independent of Sensory Processing and Behavior." *Science Translational Medicine* 5 (198): 198ra105.

Casarotto, Silvia, et al. 2016. "Stratification of Unresponsive Patients by an Independently Validated Index of Brain Complexity." *Annals of Neurology* 80 (5): 718–729.

CBS News (60 Minutes). 16 April 2023. "Google's Developers on the Future of AI (Transcript)." Accessed 2 November 2025.

Chaisson, Eric J. 2001. *Cosmic Evolution: The Rise of Complexity in Nature*. Cambridge, MA: Harvard University Press.

Chaisson, Eric J. 2011. "Energy Rate Density as a Complexity Metric and Evolutionary Driver." *Complexity* 16 (3): 27–40.

Christakis, Nicholas A., and James H. Fowler. 2009. *Connected: The Surprising Power of Our Social Networks and How They Shape Our Lives*. New York: Little, Brown and Company.

Christiano, Paul F., et al. 2017. "Deep Reinforcement Learning from Human Preferences." In *Proceedings of NeurIPS 2017*. https://arxiv.org/abs/1706.03741.

Chwe, Michael Suk-Young. 2001. *Rational Ritual: Culture, Coordination, and Common Knowledge*. Princeton, NJ: Princeton University Press.

Cialdini, Robert B., Carl A. Kallgren, and Raymond R. Reno. 1990. "A Focus Theory of Normative Conduct." In *Advances in Experimental Social Psychology*, Vol. 24, 201–234. New York: Academic Press.

Clark, Andy. 2016. *Surfing Uncertainty: Prediction, Action, and the Embodied Mind*. Oxford: Oxford University Press.

Collini, Elisabetta, Cathy Y. Wong, Krystyna E. Wilk, Paul M. G. Curmi, Paul Brumer, and Gregory D. Scholes. 2010. "Coherently Wired Light-Harvesting in Photosynthetic Marine Algae at Ambient Temperature." *Nature* 463: 644–647.

Collins, Randall. 2004. *Interaction Ritual Chains*. Princeton, NJ: Princeton University Press.

Connell, Joseph H. 1978. "Diversity in Tropical Rain Forests and Coral Reefs." *Science* 199 (4335): 1302–1310.

Conway Morris, Simon. 2003. *Life's Solution: Inevitable Humans in a Lonely Universe*. Cambridge: Cambridge University Press.

Couzin, Iain D., Jens Krause, Richard James, Graeme D. Ruxton, and Nigel R. Franks. 2002. "Collective Memory and Spatial Sorting in Animal Groups." *Journal of Theoretical Biology* 218 (1): 1–11.

Crick, Francis H. C. 1958. "On Protein Synthesis." *Symposia of the Society for Experimental Biology* 12: 138–163.

Crick, Francis H. C. 1968. "The Origin of the Genetic Code." *Journal of Molecular Biology* 38 (3): 367–379.

Crow, James F., and Motoo Kimura. 1970. *An Introduction to Population Genetics Theory*. New York: Harper & Row.

Darwin, Charles. 1859. *On the Origin of Species by Means of Natural Selection*. London: John Murray.

Davidson, Eric H. 2001. *Genomic Regulatory Systems: Development and Evolution*. San Diego: Academic Press.

Davies, Paul. 2019. *The Demon in the Machine: How Hidden Webs of Information Are Solving the Mystery of Life*. London: Allen Lane.

Dawkins, Richard. 1982. *The Extended Phenotype: The Gene as the Unit of Selection*. Oxford: Oxford University Press.

Deacon, Terrence W. 2011. *Incomplete Nature: How Mind Emerged from Matter*. New York: W. W. Norton.

Deco, Gustavo, Giulio Tononi, Mélanie Boly, and Morten L. Kringelbach. 2015. "Rethinking Segregation and Integration: Contributions of Whole-Brain Modelling." *Nature Reviews Neuroscience* 16: 430–439.

Deneubourg, Jean-Louis, Simon Goss, Nigel R. Franks, and Jacques M. Pasteels. 1989. "The Dynamics of Collective Sorting: Robot-Like Ants and Ant-Like Robots." In

From Animals to Animats: Proceedings of the First International Conference on Simulation of Adaptive Behavior, 356–363. Cambridge, MA: MIT Press.

Dennard, Robert H. 1968. "Field-Effect Transistor Memory." U.S. Patent 3,387,286, filed July 14, 1967, and issued June 4, 1968.

Dennard, Robert H., Fritz H. Gaensslen, Hwa-Nien Yu, V. Leo Rideout, Ernest Bassous, and Andre R. LeBlanc. 1974. "Design of Ion-Implanted MOSFETs with Very Small Dimensions." *IEEE Journal of Solid-State Circuits* 9 (5): 256–268.

Dennett, Daniel C. 1995. *Darwin's Dangerous Idea: Evolution and the Meanings of Life*. New York: Simon & Schuster.

Descartes, René. 1984. "Meditations on First Philosophy." In *The Philosophical Writings of Descartes*, Vol. 2, translated by John Cottingham, Robert Stoothoff, and Dugald Murdoch. Cambridge: Cambridge University Press. (Orig. pub. 1641.)

Dever, Thomas E., and Rachel Green. 2012. "The Elongation, Termination, and Recycling Phases of Translation." *Cold Spring Harbor Perspectives in Biology* 4 (7): a013706.

Devlin, Jacob, Ming-Wei Chang, Kenton Lee, and Kristina Toutanova. 2019. "BERT: Pre-Training of Deep Bidirectional Transformers for Language Understanding." In *Proceedings of NAACL-HLT 2019*. https://arxiv.org/abs/1810.04805.

Dias, Brian G., and Kerry J. Ressler. 2014. "Parental Olfactory Experience Influences Behavior and Neural Structure in Subsequent Generations." *Nature Neuroscience* 17 (1): 89–96.

Dill, Ken A. 1990. "Dominant Forces in Protein Folding." *Biochemistry* 29 (31): 7133–7155.

Dill, Ken A., and Justin L. MacCallum. 2012. "The Protein-Folding Problem, 50 Years on." *Science* 338 (6110): 1042–1046.

Dobzhansky, Theodosius. 1937. *Genetics and the Origin of Species*. New York: Columbia University Press.

Dunbar, Robin I. M. 1996. *Grooming, Gossip, and the Evolution of Language*. Cambridge, MA: Harvard University Press.

Dunbar, Robin I. M. 1998. "The Social Brain Hypothesis." *Evolutionary Anthropology* 6 (5): 178–190.

Durkheim, Émile. 1912. *The Elementary Forms of Religious Life*. New York: Free Press. (Translated 1995.)

Dyson, Frank W., Arthur S. Eddington, and Charles Davidson. 1920. "A Determination of the Deflection of Light by the Sun's Gravitational Field, from Observations Made at the Total Eclipse of May 29, 1919." *Philosophical Transactions of the Royal Society of London, Series A* 220: 291–333.

Edelman, Gerald M. 1989. *The Remembered Present: A Biological Theory of Consciousness*. New York: Basic Books.

Einstein, Albert. 1905. "Ist die Trägheit eines Körpers von seinem Energieinhalt abhängig?" *Annalen der Physik* 18: 639–641.

Einstein, Albert. 1915. "Die Feldgleichungen der Gravitation." *Sitzungsberichte der Königlich Preussischen Akademie der Wissenschaften (Berlin)*: 844–847.

Einstein, Albert, Boris Podolsky, and Nathan Rosen. 1935. "Can Quantum-Mechanical Description of Physical Reality Be Considered Complete?" *Physical Review* 47 (10): 777–780.

Eisenberger, Naomi I., Matthew D. Lieberman, and Kipling D. Williams. 2003. "Does Rejection Hurt? An fMRI Study of Social Exclusion." *Science* 302 (5643): 290–292.

Eisenstein, Elizabeth L. 1979. *The Printing Press as an Agent of Change*. 2 Vols. Cambridge: Cambridge University Press.

Eldredge, Niles, and Stephen Jay Gould. 1972. "Punctuated Equilibria: An Alternative to Phyletic Gradualism." In *Models in Paleobiology*, edited by Thomas J. M. Schopf, 82–115. San Francisco: Freeman, Cooper.

Emeneau, Murray B. 1956. "India as a Linguistic Area." *Language* 32 (1): 3–16.

Encyclopaedia Britannica. n.d. "Operation Husky." *Britannica.* Accessed November 2, 2025.

Engel, Gregory S., Tessa R. Calhoun, Elizabeth L. Read, Tae-Kyu Ahn, Tomas Mančal, Yuan-Chung Cheng, Robert E. Blankenship, and Graham R. Fleming. 2007. "Evidence for Wavelike Energy Transfer Through Quantum Coherence in Photosynthetic Systems." *Nature* 446: 782–786. https://doi.org/10.1038/nature0567.

Engelbart, Douglas C. 1962. *Augmenting Human Intellect: A Conceptual Framework.* Menlo Park, CA: SRI.

England, Jeremy L. 2013. "Statistical Physics of Self-Replication." *The Journal of Chemical Physics* 139 (12): 121923.

England, Jeremy L. 2015. "Dissipative Adaptation in Driven Self-Assembly." *Nature Nanotechnology* 10 (11): 919–923.

Engler, Adam J., Shamik Sen, H. Lee Sweeney, and Dennis E. Discher. 2006. "Matrix Elasticity Directs Stem Cell Lineage Specification." *Cell* 126 (4): 677–689.

Ertem, Gözen, and James P. Ferris. 1993. "Template-Directed Synthesis of RNA Oligomers on Mineral Surfaces." *Nature* 361 (6408): 170–172.

European Union. 4 May 2016. "Regulation (EU) 2016/679 of the European Parliament and of the Council of 27 April 2016 (General Data Protection Regulation)." *Official Journal of the European Union L* 119: 1–88.

Ewens, Warren J. 2004. *Mathematical Population Genetics 1: Theoretical Introduction.* 2nd ed. New York: Springer.

Feller, William. 1950. *An Introduction to Probability Theory and Its Applications.* Vol. 1. New York: Wiley.

Ferris, James P. 2002. "Montmorillonite Catalysis of 30–50 Mer Oligonucleotides: Laboratory Demonstration of Potential Steps in the Origin of the RNA World." *Origins of Life and Evolution of the Biosphere* 32 (4): 311–332.

Ferris, James P., and Gözde Ertem. 1993. "Montmorillonite Catalysis of RNA Oligomer Formation in Aqueous Solution: A Model for the Prebiotic Formation of RNA." *Journal of the American Chemical Society* 115: 12270–12275.

Feynman, Richard P., Robert B. Leighton, and Matthew Sands. 1965. *The Feynman Lectures on Physics. Vol. 3: Quantum Mechanics.* Reading, MA: Addison-Wesley.

Fields, R. Douglas. 2005. "Myelination: An Overlooked Mechanism of Synaptic Plasticity?" *The Neuroscientist* 11 (6): 528–531.

Fish, Frank E. 1996. "Transitions from Drag-Based to Lift-Based Propulsion in Mammalian Swimming." *American Zoologist* 36 (6): 628–641.

Fisher, Matthew P. A. 2015. "Quantum Cognition: The Possibility of Processing Information with Nuclear Spins in the Brain." *Annals of Physics* 362: 593–602.

Fisher, Ronald A. 1930. *The Genetical Theory of Natural Selection.* Oxford: Clarendon.

Fitch, W. Tecumseh. 2010. *The Evolution of Language.* Cambridge: Cambridge University Press.

Floridi, Luciano. 2011. *The Philosophy of Information.* Oxford: Oxford University Press.

Forde, Brian G., and Hector Lorenzo. 2001. "The Nutritional Control of Root Development." *Plant and Soil* 232 (1): 51–68.

Forsythe, Jay G., Sheng-Sheng Yu, Irena Mamajanov, Martha A. Grover, Ramanarayanan Krishnamurthy, Facundo M. Fernández, and Nicholas V. Hud. 2015. "Ester-Mediated Amide Bond Formation Driven by Wet–Dry Cycles: A Possible

Path to Polypeptides on the Prebiotic Earth." *Angewandte Chemie International Edition* 54 (34): 9871–9875.

Frenkel-Pinter, Meirav, et al. 2021. "Prebiotic Peptides and the RNA-Peptide World." *Chemical Reviews* 121 (4): 2434–2461.

Fries, Pascal. 2005. "A Mechanism for Cognitive Dynamics: Neuronal Communication Through Neuronal Coherence." *Trends in Cognitive Sciences* 9 (10): 474–480.

Frisch, Priscilla C., et al. 2011. "The Galactic Environment of the Sun: Interstellar Material Inside and Outside of the Heliosphere." *Annual Review of Astronomy and Astrophysics* 49: 237–279.

Friston, Karl. 2005. "A Theory of Cortical Responses." *Philosophical Transactions of the Royal Society B: Biological Sciences* 360 (1456): 815–836.

Friston, Karl. 2010. "The Free-Energy Principle: A Unified Brain Theory?" *Nature Reviews Neuroscience* 11 (2): 127–138.

Furukawa, Yoshihiro, et al. 2019. "Extraterrestrial Ribose and Other Sugars in Primitive Meteorites." *Proceedings of the National Academy of Sciences* 116 (49): 24440–24445.

Garrison, Peter. 1970s–2010s. "Essays on Airflow Optimization and Winglets." *Flying Magazine* (various issues). (Contextual commentary; cite specific articles/dates if available.)

Gehring, Walter J. 1996. "The Master Control Gene for Morphogenesis and Evolution of the Eye." *Genes to Cells* 1 (1): 11–15.

Gell-Mann, Murray. 1994. "Complex Adaptive Systems." In *Complexity: Metaphors, Models, and Reality*, edited by George A. Cowan, David Pines, and David Meltzer, 17–45. Reading, MA: Addison-Wesley.

Gibbs, J. Willard. 1902. *Elementary Principles in Statistical Mechanics*. New York: Charles Scribner's Sons.

Gleick, James. 2011. *The Information: A History, a Theory, a Flood*. New York: Pantheon.

Goffman, Erving. 1959. *The Presentation of Self in Everyday Life*. New York: Anchor.

Goffman, Erving. 1974. *Frame Analysis: An Essay on the Organization of Experience*. New York: Harper & Row.

Goldstein, Noah J., Robert B. Cialdini, and Vladas Griskevicius. 2008. "A Room with a Viewpoint: Using Social Norms to Motivate Environmental Conservation." *Journal of Consumer Research* 35 (3): 472–482.

Gordon, Deborah M. 2010. *Ant Encounters: Interaction Networks and Colony Behavior*. Princeton, NJ: Princeton University Press.

Gould, Stephen Jay. 1989. *Wonderful Life: The Burgess Shale and the Nature of History*. New York: W. W. Norton.

Gould, Stephen Jay. 1996. *Full House: The Spread of Excellence from Plato to Darwin*. New York: Harmony Books.

Grimaldi, David, and Michael S. Engel. 2005. *Evolution of the Insects*. Cambridge: Cambridge University Press.

Haidt, Jonathan. 2001. "The Emotional Dog and Its Rational Tail." *Psychological Review* 108 (4): 814–834.

Haig, David. 2002. *Genomic Imprinting and Kinship*. New Brunswick, NJ: Rutgers University Press.

Haken, Hermann. 1977. *Synergetics: An Introduction*. Berlin: Springer.

Hameroff, Stuart, and Roger Penrose. 2014. "Consciousness in the Universe: A Review of the 'Orch OR' Theory." *Physics of Life Reviews* 11 (1): 39–78.

Hanczyc, Martin M., Sergey M. Fujikawa, and Jack W. Szostak. 2003. "Experimental Models of Primitive Cellular Compartments: Encapsulation, Growth, and Division." *Science* 302 (5645): 618–622.

Hartl, F. Ulrich, and Manajit Hayer-Hartl. 2009. "Converging Concepts of Protein Folding in Vitro and in Vivo." *Nature Structural & Molecular Biology* 16 (6): 574–581.

Hawking, Stephen W. 1974. "Black Hole Explosions?" *Nature* 248: 30–31.

Hawking, Stephen W. 1975. "Particle Creation by Black Holes." *Communications in Mathematical Physics* 43 (3): 199–220.

Hawking, Stephen W. 1976. "Breakdown of Predictability in Gravitational Collapse." *Physical Review D* 14 (10): 2460–2473.

He, Christine, Isaac Gállego, Brandon Laughlin, Martha A. Grover, and Nicholas V. Hud. 2017. "A Viscous Solvent Enables Information Transfer from Gene-Length Nucleic Acids in a Model Prebiotic Replication Cycle." *Nature Chemistry* 9 (4): 318–324.

Heard, Edith, and Rob Martienssen. 2014. "Transgenerational Epigenetic Inheritance: Myths and Mechanisms." *Cell* 157 (1): 95–109.

Hebb, Donald O. 1949. *The Organization of Behavior: A Neuropsychological Theory*. New York: Wiley.

Heijmans, Bastiaan T., et al. 2008. "Persistent Epigenetic Differences Associated with Prenatal Exposure to Famine in Humans." *Proceedings of the National Academy of Sciences* 105 (44): 17046–17049.

Helbing, Dirk, Illés J. Farkas, and Tamás Vicsek. 2000. "Simulating Dynamical Features of Escape Panic." *Nature* 407: 487–490.

Helbing, Dirk, and Péter Molnár. 1995. "Social Force Model for Pedestrian Dynamics." *Physical Review E* 51 (5): 4282–4286.

Henningsson, Per, and Anders Hedenström. 2011. "Aerodynamics of Gliding Flight in a Swift." *Journal of Experimental Biology* 214 (3): 382–393.

Henrich, Joseph. 2015. *The Secret of Our Success*. Princeton, NJ: Princeton University Press.

Hensen, Bas, et al. 2015. "Loophole-free Bell Inequality Violation Using Electron Spins Separated by 1.3 Kilometres." *Nature* 526 (7575): 682–686.

Heylighen, Francis. 2007. "The Global Superorganism: An Evolutionary-Cybernetic Model of the Emerging Network Society." *Social Evolution & History* 6 (1): 58–119.

Hinnebusch, Alan G. 2014. "The Scanning Mechanism of Eukaryotic Translation Initiation." *Annual Review of Biochemistry* 83: 779–812.

Hobbes, Thomas. 1651. *Leviathan*. London: Andrew Crooke.

Hoffmeyer, Jesper. 2008. *Biosemiotics: An Examination into the Signs of Life and the Life of Signs*. Scranton, PA: University of Scranton Press.

Hogenesch, John B., and Hiroki R. Ueda. 2011. "Understanding Systems-Level Properties: Timely Stories from the Study of Clocks." *Nature Reviews Genetics* 12 (6): 407–416.

Hohmann, Walter. 1925. *Die Erreichbarkeit der Himmelskörper: Untersuchungen über das Raumfahrtproblem*. München: R. Oldenbourg.

Hohwy, Jakob. 2013. *The Predictive Mind*. Oxford: Oxford University Press.

Hopfield, John J. 1974. "Kinetic Proofreading: A New Mechanism for Reducing Errors in Biosynthetic Processes Requiring High Specificity." *Proceedings of the National Academy of Sciences* 71 (10): 4135–4139.

Hore, Peter J., and Henrik Mouritsen. 2016. "The Radical-Pair Mechanism of Magnetoreception." *Annual Review of Biophysics* 45: 299–344.

Horning, David P., and Gerald F. Joyce. 2016. "Amplification of RNA by an RNA Polymerase Ribozyme." *Proceedings of the National Academy of Sciences* 113 (35): 9786–9791.

Hubbard, E. M., and V. S. Ramachandran. 2005. "Neurocognitive Mechanisms of Synesthesia." *Neuron* 48 (3): 509–520.

Hudson, Graham A., et al. 2013. "Emergence of a Prebiotically Plausible Ribozyme That Catalyzes RNA Ligation." *Nature Chemistry* 5 (11): 1006–1014.

Huelga, Susana F., and Martin B. Plenio. 2013. "Vibrations, Quanta and Biology." *Contemporary Physics* 54 (4): 181–207.

Huttenlocher, Peter R. 1979. "Synaptic Density in Human Frontal Cortex—Developmental Changes During Aging." *Brain Research* 163 (2): 195–205.

Huxley, Julian. 1942. *Evolution: The Modern Synthesis*. London: George Allen & Unwin.

Ibba, Michael, and Dieter Söll. 2000. "Aminoacyl-tRNA Synthesis." *Annual Review of Biochemistry* 69: 617–650.

IBM Archives. n.d. "IBM 350 Disk Storage Unit." Accessed 2 November 2025. https://www.ibm.com/ibm/history/exhibits/storage/storage_350.html.

Immink, Kees A. Schouhamer. July–August 1999. "The Compact Disc Story." *Journal of the Audio Engineering Society* 46 (5): 458–465.=7

Ingolia, Nicholas T. 2011. "Genome-Wide Translational Profiling by Ribosome Footprinting." *Nature Reviews Genetics* 12 (6): 453–469.

Jablonka, Eva, and Marion J. Lamb. 2005. *Evolution in Four Dimensions: Genetic, Epigenetic, Behavioral, and Symbolic Variation in the History of Life*. Cambridge, MA: MIT Press.

Jacob, François. 1977. "Evolution and Tinkering." *Science* 196 (4295): 1161–1166.

Jacob, François, and Jacques Monod. 1961. "Genetic Regulatory Mechanisms in the Synthesis of Proteins." *Journal of Molecular Biology* 3 (3): 318–356.

Janeway, Charles A., Paul Travers, Mark Walport, and Mark J. Shlomchik. 2001. *Immunobiology: The Immune System in Health and Disease*. 5th ed. New York: Garland Science.

Jannasch, Holger W., and Michael J. Mottl. 1985. "Geochemical and Biogeochemical Interactions at Hydrothermal Vents." *Science* 229 (4715): 717–725.

Janzen, Fredric J., and Patrick C. Phillips. 2006. "Exploring the Evolution of Environmental Sex Determination, Especially in Reptiles." *Journal of Evolutionary Biology* 19 (6): 1775–1784.

Jaynes, E. T. 1957. "Information Theory and Statistical Mechanics." *Physical Review* 106 (4): 620–630; 108 (2): 171–190.

Jeffery, William R. 2001. "Cavefish as a Model System in Evolutionary Developmental Biology." *Developmental Biology* 231 (1): 1–12.

Jerison, Harry J. 1973. *Evolution of the Brain and Intelligence*. New York: Academic Press.

Jirtle, Randy L., and Michael K. Skinner. 2007. "Environmental Epigenomics and Disease Susceptibility." *Nature Reviews Genetics* 8 (4): 253–262.

Johnson, Melvin, et al. 2017. "Google's Multilingual Neural Machine Translation System: Bridging the Gap Between Human and Machine Translation." https://arxiv.org/abs/1611.04558.

Johnson, Steven. 2001. *Emergence: The Connected Lives of Ants, Brains, Cities, and Software*. New York: Scribner.

Jönsson, Claus. 1961. "Elektroneninterferenzen an mehreren künstlich hergestellten Feinspalten." *Zeitschrift für Physik* 161 (4): 454–474.

Joyce, Gerald F., and Leslie E. Orgel. 1999. "Prospects for Understanding the Origin of the RNA World." In *The RNA World*, 2nd ed., edited by Raymond F. Gesteland, Thomas R. Cech, and John F. Atkins, 49–77. Cold Spring Harbor, NY: Cold Spring Harbor Laboratory Press.

Jumper, John, et al. 2021. "Highly Accurate Protein Structure Prediction with AlphaFold." *Nature* 596: 583–589.

Kaplan, Jared, et al. 2020. "Scaling Laws for Neural Language Models." https://arxiv.org/abs/2001.08361.

Kauffman, Stuart A. 1993. *The Origins of Order: Self-Organization and Selection in Evolution*. Oxford: Oxford University Press.

Kelman, Herbert C. 1958. "Compliance, Identification, and Internalization: Three Processes of Attitude Change." *Journal of Conflict Resolution* 2 (1): 51–60.

Khinchin, A. I. 1949. *Mathematical Foundations of Statistical Mechanics*. New York: Dover.

Kimura, Motoo. 1964. "Diffusion Models in Population Genetics." *Journal of Applied Probability* 1 (2): 177–232.

Kimura, Motoo. 1968. "Evolutionary Rate at the Molecular Level." *Nature* 217 (5129): 624–626.

Kirschner, Marc W., and John C. Gerhart. 1998. "Evolvability." *Proceedings of the National Academy of Sciences of the United States of America* 95 (15): 8420–8427.

Klinman, Judith P., and Amnon Kohen. 2013. "Hydrogen Tunneling Links Protein Dynamics to Enzyme Catalysis." *Annual Review of Biochemistry* 82: 471–496.

Knapp, Michael J., Kurt Rickert, and Judith P. Klinman. 2002. "Temperature-Dependent Isotope Effects in Soybean Lipoxygenase-1: Correlating Hydrogen Tunneling with Protein Dynamics." *Journal of the American Chemical Society* 124 (15): 3865–3874.

Knight, Russell D., Stephen J. Freeland, and Laura F. Landweber. 1999. "Selection, History and Chemistry: The Three Faces of the Genetic Code." *Trends in Biochemical Sciences* 24 (6): 241–247.

Koizumi, Osamu. 2002. "Developmental Neurobiology of Hydra, a Model Animal of Cnidarians." *Canadian Journal of Zoology* 80 (10): 1678–1689.

Kolmogorov, Andrei N. 1931. "Über die analytischen Methoden in der Wahrscheinlichkeitsrechnung." *Mathematische Annalen* 104: 415–458.

Kressler, Dieter, Ed Hurt, and Jochen Bassler. 2010. "Driving Ribosome Assembly." *Nature Reviews Molecular Cell Biology* 11 (2): 112–125.

Kristan, William B., Jr., Ronald L. Calabrese, and William O. Friesen. 2005. "Neuronal Control of Leech Behavior." *Progress in Neurobiology* 76 (5): 279–327.

Lallement, Rosine, Barry Y. Welsh, Jean-Luc Vergely, Françoise Crifo, and Darri Sfeir. 2003. "3D Mapping of the Dense Interstellar Gas around the Local Bubble." *Astronomy & Astrophysics* 411: 447–464.

Lambert, Neill, Yueh-Nan Chen, Yuan-Chung Cheng, Che-Ming Li, Guang-Yin Chen, and Franco Nori. 2013. "Quantum Biology." *Nature Physics* 9: 10–18.

Land, Michael F., and Dan-Eric Nilsson. 2002. *Animal Eyes*. Oxford: Oxford University Press.

Landauer, Rolf. 1961. "Irreversibility and Heat Generation in the Computing Process." *IBM Journal of Research and Development* 5 (3): 183–191.

Landauer, Rolf. 1991. "Information Is Physical." *Physics Today* 44 (5): 23–29.

Laughlin, Robert B., and David Pines. 2000. "The Theory of Everything." *Proceedings of the National Academy of Sciences of the United States of America* 97 (1): 28–31.

LeCun, Yann, John S. Denker, and Sara A. Solla. 1990. "Optimal Brain Damage." In *Advances in Neural Information Processing Systems 2*, 598–605. San Mateo, CA: Morgan Kaufmann.

Leiner, Barry M., et al. n.d. "A Brief History of the Internet." *Internet Society*. Accessed 2 November 2025.

Leopold, Luna B., and M. Gordon Wolman. 1957. "River Channel Patterns: Braided, Meandering, and Straight." U.S. Geological Survey Professional Paper 282-B. Washington, DC: U.S. Government Printing Office.

Levin, Simon A. 1998. "Ecosystems and the Biosphere as Complex Adaptive Systems." *Ecosystems* 1 (5): 431–436.

Levinthal, Cyrus. 1969. "How to Fold Graciously." In *Mössbauer Spectroscopy in Biological Systems: Proceedings of a Meeting Held at Allerton House, Monticello, Illinois*, 22–24. Urbana, IL: University of Illinois Press.

Lewis, David. 1969. *Convention: A Philosophical Study*. Cambridge, MA: Harvard University Press.

Lewontin, Richard C. 2000. *The Triple Helix: Gene, Organism, and Environment.* Cambridge, MA: Harvard University Press.

Li, Liangjie, et al. 2017. "Enhanced Nonenzymatic RNA Copying with 2-Aminoimidazole Activated Nucleotides." *Journal of the American Chemical Society* 139 (5): 1810–1813.

Lim, James, David Paleček, Felipe Caycedo-Soler, Craig N. Lincoln, Javier Prior, Hans von Berlepsch, Susana F. Huelga, Martin B. Plenio, Donatas Zigmantas, and Jürgen Hauer. 2015. "Vibronic Origin of Long-Lived Coherence in an Artificial Molecular Light Harvester." *Nature Communications* 6: 7755.

Lineweaver, Charles H., Yeshe Fenner, and Brad K. Gibson. 2004. "The Galactic Habitable Zone and the Age Distribution of Complex Life in the Milky Way." *Science* 303 (5654): 59–62.

Liouville, Joseph. 1838. "Note on the Theory of Variable Transformations and Their Jacobians." *Journal de Mathématiques Pures et Appliquées* 3: 116–120.

Lloyd, Seth. 2006. *Programming the Universe: A Quantum Computer Scientist Takes on the Cosmos.* New York: Alfred A. Knopf.

Lovelock, James. 1979. *Gaia: A New Look at Life on Earth.* Oxford: Oxford University Press.

Ma, Xiao-Song, Johannes Kofler, and Anton Zeilinger. 2016. "Delayed-Choice Gedanken Experiments and Their Realizations." *Reviews of Modern Physics* 88 (1): 015005.

Macintyre, Ben. 2010. *Operation Mincemeat: The True Spy Story That Changed the Course of World War II.* London: Bloomsbury.

Maeda, John. 2006. *The Laws of Simplicity.* Cambridge, MA: MIT Press.

Maldacena, Juan. 1998. "The Large-N Limit of Superconformal Field Theories and Supergravity." *Advances in Theoretical and Mathematical Physics* 2 (2): 231–252.

Mandelbrot, Benoit B. 1982. *The Fractal Geometry of Nature.* San Francisco: W. H. Freeman.

Mann, Stephen. 2012. "Systems of Creation: The Emergence of Life from Nonliving Matter." *Accounts of Chemical Research* 45 (12): 2131–2141.

Mansy, Sheref S., and Jack W. Szostak. 2008. "Thermostability of Model Protocell Membranes." *Proceedings of the National Academy of Sciences* 105 (36): 13351–13356.

Mashour, George A. 2014. "Top-Down Mechanisms of Anesthetic-Induced Unconsciousness." *Frontiers in Systems Neuroscience* 8: 115.

Masuoka, Fujio, Masuoka Momodomi, Y. Iwata, and R. Shirota. 1987. "New Ultra High Density EPROM and Flash EEPROM with NAND Structure Cell." In *Technical Digest of the 1987 IEEE International Electron Devices Meeting (IEDM)*, 552–555. New York: IEEE.

Mayer-Schönberger, Viktor, and Kenneth Cukier. 2013. *Big Data: A Revolution That Will Transform How We Live, Work, and Think.* Boston: Houghton Mifflin Harcourt.

Maynard Smith, John. 2000. "The Concept of Information in Biology." *Philosophy of Science* 67 (2): 177–194.

Maynard Smith, John, Richard Burian, Stuart Kauffman, Pere Alberch, Brian Goodwin, Russell Lande, David Raup, and Leigh Van Valen. 1985. "Developmental Constraints and Evolution." *Quarterly Review of Biology* 60 (3): 265–287.

Maynard Smith, John, and Eörs Szathmáry. 1995. *The Major Transitions in Evolution.* Oxford: Oxford University Press.

Maynard Smith, John, and Eörs Szathmáry. 1999. *The Origins of Life: From the Birth of Life to the Origin of Language.* Oxford: Oxford University Press.

McComb, Karen, Lucy Moss, Sarah M. Durant, Catherine Baker, and Shannon Sayialel. 2001. "Matriarchs as Repositories of Social Knowledge in African Elephants." *Science* 292 (5516): 491–494.

McLuhan, Marshall. 1964. *Understanding Media: The Extensions of Man*. New York: McGraw-Hill.

McWhorter, John H. 2011. *What Language Is (and What It Isn't)*. New York: Gotham Books.

Mech, L. David, and Luigi Boitani, eds. 2003. *Wolves: Behavior, Ecology, and Conservation*. Chicago: University of Chicago Press.

Minkowski, Hermann. 1909. "Raum und Zeit." *Physikalische Zeitschrift* 10: 75–88. (Lecture delivered 1908 at the 80th Assembly of German Natural Scientists and Physicians, Cologne.)

Minovitch, Michael A. 1963. "The Determination and Characteristics of Ballistic Interplanetary Trajectories Under the Influence of Multiple Planetary Attractions." JPL Technical Report 32-464. Pasadena, CA: Jet Propulsion Laboratory.

Mitchell, Melanie. 2009. *Complexity: A Guided Tour*. New York: Oxford University Press.

Moore, Gordon E. 1965. "Cramming More Components onto Integrated Circuits." *Electronics* 38 (8): 114–117.

Moss, Cynthia. 1988. *Elephant Memories: Thirteen Years in the Life of an Elephant Family*. New York, NY: William Morrow.

Muchowska, Katarzyna B., Sergey J. Varma, and Joseph Moran. 2019. "Synthesis of Carboxylic Acids in Mineral-Promoted, Prebiotically Plausible Reactions." *Nature* 569 (7754): 104–107.

The National Archives (UK). n.d. "Domesday Book." Accessed 2 November 2025. https://www.nationalarchives.gov.uk/domesday/.

National People's Congress (PRC). 1 September 2021. "Data Security Law of the People's Republic of China." Accessed 2 November 2025. https://www.npc.gov.cn/englishnpc/c2759/c23934/202112/t20211209_385109.html.

National People's Congress (PRC). 1 November 2021. "Personal Information Protection Law of the People's Republic of China." Accessed 2 November 2025. https://en.npc.gov.cn/2021-12/29/c_694559.htm.

Nicolis, Grégoire, and Ilya Prigogine. 1977. *Self-Organization in Nonequilibrium Systems: From Dissipative Structures to Order Through Fluctuations*. New York: Wiley.

Niederholtmeyer, Henrike, et al. 2015. "Rapid Cell-Free Transcription-Translation from Lipid-Gated Compartments." *Nature Communications* 6: 10047.

Nielsen, Michael A., and Isaac L. Chuang. 2010. *Quantum Computation and Quantum Information*. 10th anniversary ed. Cambridge: Cambridge University Press.

Nikam, N. A., and Richard McKeon, trans. 1959. *The Edicts of Asoka*. Chicago: University of Chicago Press.

Nilsson, Dan-Eric. 2013. "Eye Evolution and Its Functional Basis." *Current Biology* 23 (17): R1016–R1022.

Nilsson, Dan-Eric, and Susanne Pelger. 1994. "A Pessimistic Estimate of the Time Required for an Eye to Evolve." *Proceedings of the Royal Society B* 256 (1345): 53–58.

Ninio, Jacques. 1975. "Kinetic Amplification of Enzyme Discrimination." *Biochimie* 57 (5): 587–595.

Nissen, Poul, Jeffrey Hansen, Joana A. Ban, Peter B. Moore, and Thomas A. Steitz. 2000. "The Structural Basis of Ribosomal Peptide Bond Formation." *Science* 289 (5481): 920–930.

Nissen, Poul, Jeffrey Hansen, Niels Ban, Peter B. Moore, and Thomas A. Steitz. 2000. "The Structural Basis of Ribosome Activity in Peptide Bond Synthesis." *Science* 289 (5481): 920–930.

Noble, Denis. 2006. *The Music of Life: Biology Beyond the Genome*. Oxford: Oxford University Press.

Norman, Donald A. 1988. *The Psychology of Everyday Things*. New York: Basic Books.

Northcutt, R. Glenn. 2002. "Understanding Vertebrate Brain Evolution." *Integrative and Comparative Biology* 42 (4): 743–756.

Ogura, Atsushi, Kazuhiro Ikeo, and Takashi Gojobori. 2004. "Comparative Analysis of Gene Expression for Vertebrate and Cephalopod Camera Eyes." *Proceedings of the National Academy of Sciences* 101 (9): 2621–2626.

Ohno, Susumu. 1970. *Evolution by Gene Duplication*. Berlin: Springer.

Oizumi, Masafumi, Larissa Albantakis, and Giulio Tononi. 2014. "From the Phenomenology to the Mechanisms of Consciousness: Integrated Information Theory 3.0." *PLoS Computational Biology* 10 (5): e1003588.

Ostrom, Elinor. 1990. *Governing the Commons: The Evolution of Institutions for Collective Action*. Cambridge, MA: Cambridge University Press.

Ouyang, Long, et al. 2022. "Training Language Models to Follow Instructions with Human Feedback." https://arxiv.org/abs/2203.02155.

Pacioli, Luca. 1494. *Summa de arithmetica, geometria, proportioni et proportionalita*. Venice: Paganino de Paganini.

Panitchayangkoon, J., D. Hayes, K. A. Fransted, J. R. Caram, E. Harel, J. Wen, R. E. Blankenship, and G. S. Engel. 2010. "Long-Lived Quantum Coherence in Photosynthetic Complexes at Physiological Temperature." *Proceedings of the National Academy of Sciences* 107 (29): 12766–12770.

Pariser, Eli. 2011. *The Filter Bubble: What the Internet Is Hiding from You*. New York: Penguin Press.

Parker, Joel, et al. 2013. "Convergent Genomic Signatures of High-Frequency Hearing in Echolocating Mammals." *Nature* 502 (7470): 228–231.

Pasquale, Frank. 2015. *The Black Box Society*. Cambridge, MA: Harvard University Press.

Patel, Bhavesh H., Claudia Percivalle, Matthew J. Ritson, Colm D. Duffy, and John D. Sutherland. 2015. "Common Origins of RNA, Protein and Lipid Precursors in a Cyanosulfidic Protosoup." *Nature Chemistry* 7 (4): 301–307.

Pecinka, Ales, and Ortrun Mittelsten Scheid. 2012. "Stress-Induced Chromatin Changes: A Critical View on Their Cause and Consequence." *Trends in Plant Science* 17 (6): 316–323.

Peebles, P. J. E. 1993. *Principles of Physical Cosmology*. Princeton, NJ: Princeton University Press.

Peres, Asher, and Daniel R. Terno. 2004. "Quantum Information and Relativity Theory." *Reviews of Modern Physics* 76 (1): 93–123.

Petrov, Alexey S., et al. 2014. "Evolution of the Ribosome at Atomic Resolution." *Proceedings of the National Academy of Sciences* 111 (28): 10251–10256.

Piantadosi, Steven T., Harry Tily, and Edward Gibson. 2011. "Word Lengths Are Optimized for Efficient Communication." *Proceedings of the National Academy of Sciences* 108 (9): 3526–3529.

Polanyi, Michael. 1958. *Personal Knowledge: Towards a Post-Critical Philosophy*. Chicago: University of Chicago Press.

Pontryagin, Lev S., V. G. Boltyanskii, R. V. Gamkrelidze, and E. F. Mishchenko. 1962. *The Mathematical Theory of Optimal Processes*. Translated by K. N. Trirogoff. New York: Interscience.

Price, George R. 1970. "Selection and Covariance." *Nature* 227 (5257): 520–521.

Prigogine, Ilya. 1967. *Introduction to Thermodynamics of Irreversible Processes*. 3rd ed. New York: Wiley.

Prigogine, Ilya. 8 December 1977. "Time, Structure and Fluctuations." Nobel Lecture. Stockholm: Nobel Foundation.

Prigogine, Ilya. 1978. "Time, Structure and Fluctuations." *Science* 201 (4358): 777–785.

Proud, Christopher G. 2019. "Phosphorylation and Signal Transduction Pathways in Translational Control." *Cold Spring Harbor Perspectives in Biology* 11 (4): a033050.

Radford, Alec, Jeffrey Wu, Rewon Child, David Luan, Dario Amodei, and Ilya Sutskever. 2019. "Language Models Are Unsupervised Multitask Learners." OpenAI Technical Report. https://cdn.openai.com/better-language-models/language_models_are_unsupervised_multitask_learners.pdf.

Radhakrishnan, S. 1927. *The Hindu View of Life*. London: George Allen & Unwin.

Raichle, Marcus E., and Deborah A. Gusnard. 2002. "Appraising the Brain's Energy Budget." *Proceedings of the National Academy of Sciences* 99 (16): 10237–10239.

Reppert, Steven M., and David R. Weaver. 2002. "Coordination of Circadian Timing in Mammals." *Nature* 418 (6901): 935–941.

Ritz, Thorsten, Peter Thalau, John B. Phillips, Roswitha Wiltschko, and Wolfgang Wiltschko. 2004. "Resonance Effects Indicate a Radical-Pair Mechanism for Avian Magnetic Compass." *Nature* 429: 177–180.

Rodnina, Marina V. 2011. "Translocation on the Ribosome: Mechanism and Kinetics." *Annual Review of Biochemistry* 80: 415–439.

Rodnina, Marina V. 2018. "Translation in Prokaryotes." *Cold Spring Harbor Perspectives in Biology* 10 (9): a032664.

Roth, Martha T., trans. 1997. *Law Collections from Mesopotamia and Asia Minor*. 2nd ed. Atlanta: Society of Biblical Literature.

Saenger, Wolfram. 1984. *Principles of Nucleic Acid Structure*. New York: Springer.

Saito, Hiroaki, Spiros Kourouklis, and Hiroaki Suga. 2001. "An RNA Enzyme That Selectively Aminoacylates a tRNA Bearing a Long Extra Arm." *Nature Structural Biology* 8 (7): 713–717.

Samanta, Biswajit, and Gerald F. Joyce. 2017. "A Reverse Transcriptase Ribozyme." *eLife* 6: e31153.

Sapir, Edward. 1921. *Language: An Introduction to the Study of Speech*. New York: Harcourt, Brace and Company.

Sarasso, Simone, et al. 2015. "Consciousness and Complexity During Unresponsiveness Induced by Propofol, Xenon, and Ketamine." *Current Biology* 25 (23): 3099–3105.

Schlichting, Carl D., and Matthew A. Wund. 2014. "Phenotypic Plasticity and Epigenetic Marking: An Assessment of Evidence for Genetic Accommodation." *Evolution* 68 (3): 656–672.

Schmeing, Thomas M., and V. Ramakrishnan. 2009. "What Recent Ribosome Structures Have Revealed About the Mechanism of Translation." *Cold Spring Harbor Perspectives in Biology* 1 (3): a003780.

Schmeing, Thomas M., Robert M. Voorhees, Amedee C. Kelley, and V. Ramakrishnan. 2005. "How Ribosomes Catalyse Peptide Bond Formation." *Molecular Cell* 20 (5): 723–732.

Schneider, Eric D., and James J. Kay. 1994. "Life as a Manifestation of the Second Law of Thermodynamics." *Mathematical and Computer Modelling* 19 (6–8): 25–48.

Schneider, Eric D., and Dorion Sagan. 2005. *Into the Cool: Energy Flow, Thermodynamics, and Life*. Chicago: University of Chicago Press.

Schneider, Thomas D. 1991. "Theory of Molecular Machines. I. Channel Capacity of Molecular Machines." *Journal of Theoretical Biology* 148 (1): 83–123.

Scholes, Gregory D., et al. 2017. "Using Coherence to Enhance Function in Chemical and Biophysical Systems." *Nature* 543: 647–656.
Schrödinger, Erwin. 1935. "Die gegenwärtige Situation in der Quantenmechanik." *Naturwissenschaften* 23: 807–812, 823–828, 844–849.
Schrödinger, Erwin. 1944. *What Is Life? The Physical Aspect of the Living Cell.* Cambridge: Cambridge University Press.
Scott, James C. 1998. *Seeing Like a State: How Certain Schemes to Improve the Human Condition Have Failed.* New Haven, CT, and London: Yale University Press.
Sennrich, Rico, Barry Haddow, and Alexandra Birch. 2016. "Neural Machine Translation of Rare Words with Subword Units." In *Proceedings of the 54th Annual Meeting of the Association for Computational Linguistics (Volume 1: Long Papers)*, 1715–1725. Stroudsburg, PA: Association for Computational Linguistics.
Seth, Anil K. 2016. "From Consciousness to Cognition and Back Again." *Trends in Cognitive Sciences* 20 (7): 365–366.
Shajani, Zahra, Michael T. Sykes, and James R. Williamson. 2011. "Assembly of Bacterial Ribosomes." *Annual Review of Biochemistry* 80: 501–526.
Shalm, Lynden K., et al. 2015. "Strong Loophole-Free Test of Local Realism." *Physical Review Letters* 115 (25): 250402.
Shannon, Claude E. July 1948 and October 1948. "A Mathematical Theory of Communication." *Bell System Technical Journal* 27 (3): 379–423; 27 (4): 623–656.
Shapiro, James A. 2011. *Evolution: A View from the 21st Century.* Upper Saddle River, NJ: Pearson FT Press.
Sherif, Muzafer. 1936. *The Psychology of Social Norms.* New York and London: Harper & Brothers.
Shoemaker, Christopher J., and Rachel Green. 2012. "Translation Drives mRNA Quality Control." *Nature Structural & Molecular Biology* 19 (6): 594–601.
Siclari, Francesca, et al. 2017. "The Neural Correlates of Dreaming." *Nature Neuroscience* 20: 872–878.
Singer, Seymour J., and Garth L. Nicolson. 1972. "The Fluid Mosaic Model of the Structure of Cell Membranes." *Science* 175 (4023): 720–731.
Skinner, Michael K., Carlos Guerrero-Bosagna, M. Muksitul Haque, Eric E. Nilsson, Jennifer A. H. Koop, Sarah A. Knutie, and Dale H. Clayton. 2014. "Epigenetics and the Evolution of Darwin's Finches." *Genome Biology and Evolution* 6 (8): 1972–1989.
Sonenberg, Nahum, and Alan G. Hinnebusch. 2009. "Regulation of Translation Initiation in Eukaryotes: Mechanisms and Biological Targets." *Cell* 136 (4): 731–745.
Starbird, Kate, Grace Muzny, and Leysia Palen. April 2012. "Learning from the Crowd: Collaborative Filtering Techniques for Identifying On-the-Ground Twitterers During Mass Disruptions." Proceedings of the 9th International ISCRAM Conference, Vancouver, Canada, 1–10.
Sterelny, Kim. 2012. *The Evolved Apprentice: How Evolution Made Humans Unique.* Cambridge, MA: MIT Press.
Stout, Dietrich, Nicholas Toth, Kathy Schick, and Thierry Chaminade. 2008. "Neural Correlates of Early Stone Age Toolmaking: Technology, Language and Cognition in Human Evolution." *Philosophical Transactions of the Royal Society B: Biological Sciences* 363 (1499): 1939–1949.
Strogatz, Steven H. 2003. *Sync: The Emerging Science of Spontaneous Order.* New York: Hyperion.
Sumpter, David J. T. 2010. *Collective Animal Behavior.* Princeton, NJ: Princeton University Press.
Susskind, Leonard. 1995. "The World as a Hologram." *Journal of Mathematical Physics* 36 (11): 6377–6396.

Susskind, Leonard, Larus Thorlacius, and John Uglum. 1993. "The Stretched Horizon and Black Hole Complementarity." *Physical Review D* 48 (8): 3743–3761.

Takahashi, Joseph S. 2017. "Transcriptional Architecture of the Mammalian Circadian Clock." *Nature Reviews Genetics* 18 (3): 164–179.

Tangney, June P., Ronda L. Stuewig, and Debra J. Mashek. 2007. "Moral Emotions and Moral Behavior." *Annual Review of Psychology* 58: 345–372.

Tegmark, Max. 2000. "The Importance of Quantum Decoherence in Brain Processes." *Physical Review E* 61 (4): 4194–4206.

Tegmark, Max. 2014. *Our Mathematical Universe: My Quest for the Ultimate Nature of Reality*. New York: Alfred A. Knopf.

't Hooft, Gerard. 1993. "Dimensional Reduction in Quantum Gravity." In *Salamfestschrift: A Collection of Talks*, edited by A. Ali, J. Ellis, and S. Randjbar-Daemi, 284–296. Singapore: World Scientific. arXiv preprint gr-qc/9310026.

Theraulaz, Guy, and Eric Bonabeau. 1999. "A Brief History of Stigmergy." *Artificial Life* 5 (2): 97–116.

Thewissen, J. G. M., E. M. Williams, L. J. Roe, and S. T. Hussain. 2001. "Skeletons of Terrestrial Cetaceans and the Relationship of Whales to Artiodactyls." *Nature* 413 (6853): 277–281.

Tjhung, Katrina F., Maxim N. Shokhirev, David P. Horning, and Gerald F. Joyce. 2020. "An RNA Polymerase Ribozyme That Synthesizes Its Own Ancestor." *Proceedings of the National Academy of Sciences of the United States of America* 117 (6): 2906–2913.

Tomasello, Michael. 1999. *The Cultural Origins of Human Cognition*. Cambridge, MA: Harvard University Press.

Tomasello, Michael. 2008. *Origins of Human Communication*. Cambridge, MA: MIT Press.

Tonomura, Akira, et al. 1989. "Demonstration of Single-Electron Buildup of an Interference Pattern." *American Journal of Physics* 57 (2): 117–120.

Tononi, Giulio. 2004. "An Information Integration Theory of Consciousness." *BMC Neuroscience* 5: 42.

Tononi, Giulio. 2008. "Consciousness as Integrated Information: A Provisional Manifesto." *The Biological Bulletin* 215 (3): 216–242.

Tononi, Giulio, Melanie Boly, Marcello Massimini, and Christof Koch. 2016. "Integrated Information Theory: From Consciousness to Its Physical Substrate." *Nature Reviews Neuroscience* 17: 450–461.

Tononi, Giulio, and Christof Koch. 2015. "Consciousness: Here, There and Everywhere?" *Philosophical Transactions of the Royal Society B: Biological Sciences* 370: 20140167.

Trudgill, Peter. 2011. *Sociolinguistic Typology: Social Determinants of Linguistic Complexity*. Oxford: Oxford University Press.

Tsien, Tsuen-Hsuin. 1985. *Paper and Printing. Vol. 5, Part 1 of Science and Civilisation in China*, edited by Joseph Needham. Cambridge: Cambridge University Press.

Tucker, Vance A. 1993. "Gliding Birds: Reduction of Induced Drag by Wingtip Slots Between the Primary Feathers." *Journal of Experimental Biology* 180 (1): 285–310.

Tufekci, Zeynep. 2013. "Not This One: Social Movements, the Attention Economy, and Microcelebrity Networked Activism." *American Behavioral Scientist* 57 (7): 848–870.

Tufekci, Zeynep. 2015. "Algorithmic Harms Beyond Facebook and Google: Emergent Challenges of Computational Agency." *Colorado Technology Law Journal* 13: 203–218.

Tully, R. Brent, Hélène Courtois, Yehuda Hoffman, and Daniel Pomarède. 2014. "The Laniakea Supercluster of Galaxies." *Nature* 513 (7516): 71–73.

Turing, Alan M. 1952. "The Chemical Basis of Morphogenesis." *Philosophical Transactions of the Royal Society of London, Series B* 237 (641): 37–72.

Turk, Rebecca M., Daniel P. Chumachenko, and Michael Yarus. 2010. "Multiple Translation Products from a Five-Base Ribozyme." *Proceedings of the National Academy of Sciences* 107 (10): 4585–4589.

Tyler, Tom R. 1990. *Why People Obey the Law*. New Haven, CT: Yale University Press.

Tyler, Tom R., and Yuen J. Huo. 2002. *Trust in the Law: Encouraging Public Cooperation with the Police and Courts*. New York: Russell Sage Foundation.

U.S. Census Bureau. n.d. "Herman Hollerith: The 1890 Census." Accessed 2 November 2025. https://www.census.gov/history/www/census_then_now/technology/hollerith_punched_card_system.html.

U.S. Department of Justice. n.d. "The CLOUD Act: Clarifying Lawful Overseas Use of Data." Accessed 2 November 2025. https://www.justice.gov/dag/cloudact.

Varela, Francisco J. 1996. "Neurophenomenology: A Methodological Remedy for the Hard Problem." *Journal of Consciousness Studies* 3 (4): 330–349.

Varela, Francisco J., Evan Thompson, and Eleanor Rosch. 1991. *The Embodied Mind: Cognitive Science and Human Experience*. Cambridge, MA: The MIT Press.

Vaswani, Ashish, Noam Shazeer, Niki Parmar, Jakob Uszkoreit, Llion Jones, Aidan N. Gomez, Łukasz Kaiser, and Illia Polosukhin. 2017. "Attention Is All You Need." In *Proceedings of the 31st Conference on Advances in Neural Information Processing Systems 30*, 5998–6008. Red Hook, NY: Curran Associates, Inc. https://arxiv.org/abs/1706.03762.

Vicsek, Tamás, András Czirók, Eshel Ben-Jacob, Inon Cohen, and Ofer Shochet. 1995. "Novel Type of Phase Transition in a System of Self-Driven Particles." *Physical Review Letters* 75 (6): 1226–1229.

von Frisch, Karl. 1967. *The Dance Language and Orientation of Bees*. Cambridge, MA: Harvard University Press.

Voorhees, Robert M., and V. Ramakrishnan. 2013. "Structural Basis of the Translational Elongation Cycle." *Annual Review of Biochemistry* 82: 203–236.

Waddington, C. H. 1942. "Canalization of Development and the Inheritance of Acquired Characters." *Nature* 150 (3811): 563–565.

Waddington, C. H. 1957. *The Strategy of the Genes*. London: Allen & Unwin.

Wagner, Andreas. 2014. *Arrival of the Fittest: Solving Evolution's Greatest Puzzle*. New York: Penguin.

Wald, George. 1958. "Foreword." In *The Fitness of the Environment*, edited by L. J. Henderson. Boston: Beacon Press.

Walton, Timothy, et al. 2020. "Mechanism of Nonenzymatic RNA Primer Extension by Imidazolium-Bridged Intermediates." *Journal of the American Chemical Society* 142 (15): 6696–6709.

Ward, Jamie. 2013. "Synesthesia." *Annual Review of Psychology* 64: 49–75.

Waters, Christopher M., and Bonnie L. Bassler. 2005. "Quorum Sensing: Cell-to-Cell Communication in Bacteria." *Annual Review of Cell and Developmental Biology* 21: 319–346.

Watson, James D., and Francis H. C. Crick. 1953. "Molecular Structure of Nucleic Acids: A Structure for Deoxyribose Nucleic Acid." *Nature* 171 (4356): 737–738.

Watts, Duncan J. 2003. *Six Degrees: The Science of a Connected Age*. New York: W. W. Norton.

Watts, Duncan J., and Steven H. Strogatz. 1998. "Collective Dynamics of 'Small-World' Networks." *Nature* 393 (6684): 440–442.

Wei, Jason, et al. 2022. "Emergent Abilities of Large Language Models." https://arxiv.org/abs/2206.07682.

West-Eberhard, Mary Jane. 2003. *Developmental Plasticity and Evolution*. Oxford: Oxford University Press.

West-Eberhard, Mary Jane. 2005. "Developmental Plasticity and the Origin of Species Differences." *Proceedings of the National Academy of Sciences* 102 (suppl. 1): 6543–6549.

Westheimer, Frank H. 1987. "Why Nature Chose Phosphates." *Science* 235 (4793): 1173–1178.

Wheeler, John Archibald. 1989. "Information, Physics, Quantum: The Search for Links." In *Proceedings of the 3rd International Symposium on Foundations of Quantum Mechanics in the Light of New Technology, Tokyo*, 354–368. Tokyo: The Physical Society of Japan.

Wheeler, John Archibald. 1990. "Information, Physics, Quantum: The Search for Links." In *Complexity, Entropy, and the Physics of Information*, edited by Wojciech H. Zurek, Santa Fe Institute Studies in the Sciences of Complexity. Redwood City, CA: Addison-Wesley.

Whitcomb, Richard T. 1976. *A Design Approach and Selected Wind-Tunnel Results at High Subsonic Speeds for Wing-Tip Mounted Winglets*. NASA Technical Note D-8260. Washington, DC: NASA.

White, Tom. 2015. *Hadoop: The Definitive Guide*. 4th ed. Sebastopol, CA: O'Reilly Media.

Wiener, Norbert. 1948. *Cybernetics: Or Control and Communication in the Animal and the Machine*. Cambridge, MA: MIT Press, Wiley.

Wilson, Edward O. 1975. *Sociobiology: The New Synthesis*. Cambridge, MA: Harvard University Press.

Wistow, Graeme, and Joram Piatigorsky. 1988. "Recruiting Proteins for Ocular Functions: The 'Gene Sharing' Concept." *Science* 240 (4851): 1300–1302.

Wochner, Alexander, James Attwater, Andrea Coulson, and Philipp Holliger. 2011. "Ribozyme-Catalyzed Transcription of an Active Ribozyme." *Science* 332 (6026): 209–212.

Wood, Wendy, and Dennis Rünger. 2016. "Psychology of Habit." *Annual Review of Psychology* 67: 289–314.

Wolfram, Stephen. 2002. *A New Kind of Science*. Champaign, IL: Wolfram Media.

Wong, J. Tze-Fei. 2005. "Coevolution Theory of the Genetic Code at Age Thirty." *BioEssays* 27 (4): 416–425.

Wright, Sewall. 1931. "Evolution in Mendelian Populations." *Genetics* 16 (2): 97–159.

Xu, Jie, et al. 2020. "Selective Prebiotic Formation of Deoxyribonucleosides by Photoredox Chemistry." *Nature* 582 (7810): 60–65.

Xu, Jingjing, et al. 2021. "Magnetic Sensitivity of Cryptochrome 4 from a Migratory Songbird." *Nature* 594: 535–540.

Young, Thomas. 1804. "The Bakerian Lecture: Experiments and Calculations Relative to Physical Optics." *Philosophical Transactions of the Royal Society of London* 94: 1–16.

Yusupov, Marat M., et al. 2001. "Crystal Structure of the Ribosome at 5.5 Å Resolution." *Science* 292 (5518): 883–896.

Zaharia, Matei, et al. 2016. "Apache Spark: A Unified Engine for Big Data Processing." *Communications of the ACM* 59 (11): 56–65.

Zaher, Hani S., and Rachel Green. 2009. "Fidelity at the Molecular Level: Lessons from Protein Synthesis." *Cell* 136 (4): 746–762.

Zahnle, Kevin J., and David C. Catling. 2017. "The Cosmic Shoreline: The Evidence That Escape Determines Which Planets Have Atmospheres." *Philosophical Transactions of the Royal Society A* 375 (2109): 20160189.

Zhang, Nai, Jack W. Szostak, and Li Li. 2022. "Kinetics of Nonenzymatic, Template-Directed RNA Primer Extension." *Journal of the American Chemical Society* 144 (10): 4518–4528.

Zilber-Rosenberg, Ilana, and Eugene Rosenberg. 2008. "Role of Microorganisms in the Evolution of Animals and Plants: The Hologenome Theory of Evolution." *FEMS Microbiology Reviews* 32 (5): 723–735.

Zipf, George Kingsley. 1949. *Human Behavior and the Principle of Least Effort*. Cambridge, MA: Addison-Wesley.

Zurek, Wojciech H. 2003. "Decoherence, Einselection, and the Quantum Origins of the Classical." *Reviews of Modern Physics* 75 (3): 715–775.

INDEX

Note: Page numbers in *italics* indicate figures.